BC next page

WELDING PROCESSES

Ivan H. Griffin
•
Edward M. Roden
•
Charles W. Briggs

 VAN NOSTRAND REINHOLD COMPANY
NEW YORK CINCINNATI TORONTO LONDON MELBOURNE

Published in 1979 by Van Nostrand Reinhold Company
A division of Litton Educational Publishing, Inc.
135 West 50th Street, New York, NY 10020, U.S.A.

Van Nostrand Reinhold Limited
1410 Birchmount Road
Scarborough, Ontario MIP 2E7, Canada

Van Nostrand Reinhold Australia Pty. Ltd.
17 Queen Street
Mitcham, Victoria 3132, Australia

Van Nostrand Reinhold Company Limited
Molly Millars Lane
Wokingham, Berkshire, England

16 15 14 13 12 11 10 9 8 7 6 5 4 3 2 1

Library of Congress Cataloging in Publication Data

Griffin, Ivan H
    Welding processes.

    Includes index.
    1. Welding.   I.   Roden, Edward M., joint author.
II.   Briggs, Charles W., joint author.   III.   Title.
TS227.G794   1979   671.5'2   78-11477
ISBN 0-442-22867-8

PREFACE

Second Edition

Welding technology, has in recent years, expanded at a fantastic rate. New ideas, processes, and techniques have arisen with such speed, that keeping abreast has become increasingly difficult. The widespread use of various methods, such as Oxyacetylene, Arc, and TIG welding in the manufacturing, construction, agriculture, and service industries, illustrates the need for this type of information. Employment in this field is highly desirable and may be secured in one of the many related areas. Therefore, pre-employment training to develop the knowledge and skills required is necessary in most cases.

WELDING PROCESSES emphasizes performance of the most commonly used methods of welding. The instructional pattern employed is one of learning by doing. Reader activities, experimentation, and testing are described and illustrated in relation to the performance of their assignments. Consideration is given throughout this book to basic principles of welding and their application to performance.

This book is directed to those who wish to improve their skills by gaining an overview of common welding techniques and related information.

The first three sections of this book are devoted to performance and principles of the most common forms of welding. The activities are suitable for learning welding as a separate occupation or as a part of another occupation, such as automobile mechanics.

The fourth section is concerned with metallurgy. This section is for those who wish to obtain a greater depth of understanding of the total welding process. To do this, numerous activities and simple experiments are provided.

The fifth section, on employment opportunities, should be of particular use to the young adult or anyone who is considering a future in the field of welding. Many job possibilities are listed and explored in this unit.

The final section contains plans and illustrations of useful projects which you can construct. The emphasis here is not only on skill development but also on design, planning, and breadth of your development.

Throughout the book, careful consideration is paid to safety practices which are incorporated in the various units in preference to a separate section on safety. The precautions should be noted.

WELDING PROCESSES has been revised to ensure thorough study of the common forms of welding. This book strives to help you attain the skills necessary to perform the various welding operations included. Therefore, the readability and content organization has been reviewed to develop a smoothness that facilitates comprehension. The technical information and the illustrations have been updated to modernize the book. A number of new projects have also been included to allow you to construct a product with the information acquired.

This book should provide the foundation upon which advanced welding techniques may be developed.

Welding technology, has in recent years, expanded at a fantastic rate. New ideas, processes, and techniques have arisen with such speed, that keeping abreast has become increasingly difficult. The widespread use of various methods, such as Oxyacetylene, Arc, and TIG welding in the manufacturing, construction, agriculture, and service industries, illustrates the need for this type of information. Employment in this field is highly desirable and may be secured in one of the many related areas. Therefore, pre-employment training to develop the knowledge and skills required is necessary in most cases.

WELDING PROCESSES emphasizes performance of the most commonly used methods of welding. The instructional pattern employed is one of learning by doing. Student activities, experimentation, and testing are described and illustrated in relation to the performance of their assignments. Consideration is given throughout this book to basic principles of welding and their application to performance.

This text is directed toward students enrolled in high school industrial arts programs and welders who wish to improve their skills by gaining an overview of common welding techniques and related information.

The first three sections of this book are devoted to performance and principles of the most common forms of welding. The activities are suitable for teaching welding as a separate occupation or as a part of another occupation, such as automobile mechanics.

The fourth section is concerned with metallurgy. This section is for those who wish to obtain a greater depth of understanding of the total welding process. To do this, numerous activities and simple experiments are provided.

The fifth section, on employment opportunities, should be of particular use to both the young adult and the student who is considering a future in the field of welding. Many job possibilities are listed and explored in this unit.

The final section contains plans and illustrations of useful projects which the student welder can construct. The emphasis here is not only on skill development but also on design, planning, and breadth of student development.

Throughout the book, careful consideration is paid to safety practices which are incorporated in the various units in preference to a separate section on safety. The precautions should be noted.

WELDING PROCESSES has been revised to ensure thorough study of the common forms of welding. This textbook strives to help the student attain the skills necessary to perform the various welding operations included. Therefore, the readability and content organization has been reviewed to develop a smoothness that facilitates comprehension. The technical information and the illustrations have been updated to modernize the book. A number of new projects have also been included to allow the student to construct a product with the information acquired.

This textbook, along with its companion texts, should provide the foundation upon which advanced welding techniques may be developed.

Other texts in this series include:

BASIC OXYACETYLENE WELDING   PIPE WELDING TECHNIQUES
BASIC ELECTRIC ARC WELDING   BLUEPRINT READING FOR WELDERS
BASIC MIG AND TIG WELDING

iv

ACKNOWLEDGMENTS

The authors wish to express their appreciation and acknowledge the contributions of the following organizations for their assistance in the development of this text:

CONTRIBUTORS

Airco Welding Products, Union, NJ 07083
Allegheny Ludlum Steel Corporation, Pittsburgh, PA 15222
Aluminum Company of America, Pittsburgh, PA 15219
American Welding Society, Miami, FL 33125
Bausch and Lomb, Rochester, NY 14602
Detroit Testing Machine Company, Detroit, MI 48213
Dow Metal Products Company, Midland, MI 48640
Hobart Brothers Company, Troy, OH 45373
Lincoln Electric Company, Cleveland, OH 44117
Linde Company, Div. of Union Carbide Corp., New York, NY 10017
Miller Electric Company, Appleton, WI 54911
Niagara Mohawk Power Corporation, Syracuse, NY 13202
Norton Company, Worcester, MA 01616
Revere Copper and Brass, Inc., New York, NY 10017
Rockingham Community College, Wentworth, NC 27375
Smith Welding Equipment, Division of Tescom Corporation, Minneapolis, MN 55441
Sylvania Electric Products Company, Towanda, PA 18848
Tempil Corporation, New York, NY 10011
United States Steel Corporation, Pittsburgh, PA 15230
Victor Equipment Company, Denton, TX 76203
Welding Design and Fabrication, Cleveland, OH 44113
Wilson Instrument, Div. American Chain and Cable, New York, NY 10017

This material has been used in the classroom by the Oswego County Board of Cooperative Educational Services, Mexico, New York. Improvements in the text have in part been a result of feedback from these classes.

CONTENTS

SECTION 3 TIG AND MIG WELDING

SECTION 4 METALLURGY AND THE WELDING INDUSTRY

SECTION 5 EMPLOYMENT OPPORTUNITIES IN WELDING

SECTION 6 PROJECTS

SECTION 1
Oxyacetylene Welding

Oxyacetylene (sometimes called gas) welding joins metal by melting them together by the heat of a torch. The general process became feasible shortly before 1900 as a result of discoveries and inventions in the latter part of the 1800s.

This type of welding is particularly useful when it is necessary for the equipment to be portable, when welding relatively thin materials, or when the amount of heat applied must be varied and controlled. Applications are made in a large number of manufacturing, construction and service industries. Two major branches of the oxyacetylene process — brazing and flame cutting — are covered in this section.

Oxyacetylene welding is performed not only by welding specialists but also by craftsmen in other occupational categories such as auto mechanics, plumbers, and the construction trades. The varied uses of oxyacetylene welding have promoted its popularity. Therefore, it is one of the most common forms of welding.

UNIT 1   THE OXYACETYLENE WELDING PROCESS

Oxyacetylene welding is one of the three basic nonpressure processes of joining metals by *fusion* alone. The process of joining two pieces by partially melting their surfaces and allowing them to flow together is called fusion. The other two fusion processes are electric arc welding and thermit welding.   Each of the three types has advantages and disadvantages.

In the oxyacetylene process, the metal is heated by the hot flame of a gas-fed torch. The metal melts and fuses together to produce the weld. In many cases, additional metal from a welding rod is melted into the joint which becomes as strong as the base metal.

EQUIPMENT AND SAFETY

The basic equipment and materials for welding by this process are:

1. Oxygen and acetylene gas supplied from cylinders to provide the flame.

2. Regulators and valves to control the flow of the gases.

3. Gages to measure the pressure of the gases.

4. Hoses to carry the gases to the torch.

5. A torch to mix the gases and to provide a handle for directing the flame.

6. A tip for the torch to control the flame.

The above equipment is described in some detail in the next several units. A thorough understanding of the equipment is highly important so that welding may be done safely as well as efficiently.   The hazards arising from lack of understanding and improper use of the equipment are:

1. Burns to the operator or nearby persons.

2. Fires in buildings or materials.

3. Explosions resulting in personal injury and property damage.

4. Damage to expensive welding equipment.

ADVANTAGES AND DISADVANTAGES

Oxyacetylene welding, brazing, and soldering operations, which are carried out with similar equipment, have certain advantages and disadvantages.

1. Oxyacetylene welding is a process which can be applied to a wide variety of manufacturing and maintenance situations.

2. The equipment is portable.

3. The cost and maintenance of the welding equipment is low when compared to that of some other welding processes.

4. The cost of welding gases, supplies, and operator's time, depends on the material being joined and the size, shape, and position in which the weld must be made.

5. The rate of heating and cooling is relatively slow. In some cases, this is an advantage. In other cases where a rapid heating and cooling cycle is desirable, the oxyacetylene welding process is not suitable.

6. A skilled operator can control the amount of heat supplied to the joint being welded. This is always a distinct advantage.

7. The oxygen and nitrogen in the air are kept from combining with the metal to form harmful oxides and nitrides.

In general, the oxyacetylene process can be used to advantage in the following situations:

- When the materials being joined are thin;

- When excessively high temperatures, or rapid heating and cooling of the work would produce unwanted or harmful changes in the metal.

- When extremely high temperatures would cause certain elements in the metal to escape into the atmosphere.

HAZARDS

Many of the hazards in oxyacetylene welding can be minimized by careful consideration of the following points:

1. The welding flame and the sparks coming from the molten puddle can cause any flammable material to ignite on contact. Therefore,

    - Flame-resistant clothing must be worn by the operator and his hair must be protected.

    - Welding and cutting should not be done near flammable materials such as wood, oil, waste or cleaning rags.

2. In addition to the risk of eye injury from flying molten metal, there is also the danger of radiation burns due to the infrared rays given off by red hot metal. The eyes may be burned if these rays are not filtered out by proper lenses. Therefore,

    - The eyes should be protected at all times by approved safety glasses and the proper shield.

    - Sunglasses are not adequate for this purpose.

3. Fluxes used in certain welding and brazing operations produce fumes which are irritating to the eyes, nose, throat, and lungs. Likewise, the fumes produced by overheating lead, zinc, and cadmium are a definite health hazard when inhaled even in small quantities. The oxides produced by these elements are poisonous. Therefore,

    - Welding should be done in a well-ventilated area.

    - The operator should not expose others to fumes produced by welding.

REVIEW QUESTIONS

1. What is the difference between fusion welding and other welding processes?

2. What features of the oxyacetylene welding process make it useful in a wide variety of jobs?

3. What are four distinct hazards that must be guarded against when oxyacetylene welding?

4. What type of rays come from the oxyacetylene flame and red hot metal?

5. What is the function of a regulator?

UNIT 2 OXYGEN AND ACETYLENE CYLINDERS

Hazards are always present when gases are compressed, stored, transported, and used under very high pressures. Oxygen and acetylene, are delivered to the user under high pressure in steel cylinders. These cylinders are made to rigid specifications.

A simple demonstration of the effects of compressing gas can be shown with an ordinary toy balloon. When the balloon is blown up and held tightly at the neck so the air cannot escape, it resembles the compressed gas cylinder. What happens if the balloon is punctured, heated, or suddenly released? The explosive burst when punctured or heated, or the sudden flight of the balloon when released shows that compressed gas, even the small amount in a toy balloon, has considerable force.

OXYGEN CYLINDERS

The most common size of oxygen cylinder, when fully charged with gas, contains 244 cubic feet of oxygen. This oxygen is at a pressure of 2,200 pounds per square inch when the temperature is 70 degrees F. (21 degrees C).

The steel walls of these cylinders are only slightly more than one-quarter inch thick, .260 inch. Dropping such a cylinder, hitting it with heavy or sharp tools, or striking an electric arc on it can cause the cylinder to explode with enough force to cause serious injury and death.

The general size and shape of an oxygen cylinder is indicated in figure 2-1. As a safety precaution, the cylinder valve is protected by a removable steel cap. This cap must be on the cylinder at all times when it is being stored or transported. The cylinder valve should always be closed when not in use, even when the cylinder is empty.

The oxygen cylinder valve is designed to handle the highly compressed oxygen gas safely. The essential parts of the valve are shown in figure 2-2. The threads on the nozzle must be protected at all times.

The *bursting disc* and *safety cap* are designed to allow the gas in the cylinder to escape if the cylinder is subjected to undue heat and the pressure in the tank begins to rise.

The double-seating valve is designed to seal off any oxygen that might leak around the valve stem. When the valve is fully open there is no leakage.

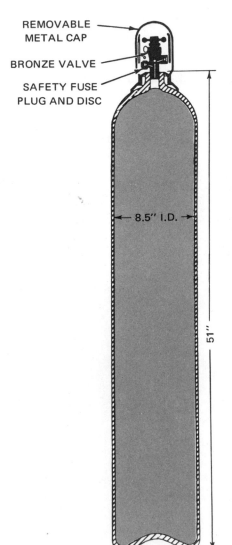

REMOVABLE METAL CAP

BRONZE VALVE

SAFETY FUSE PLUG AND DISC

← 8.5" I.D. →

51"

Fig. 2-1 Oxygen cylinder

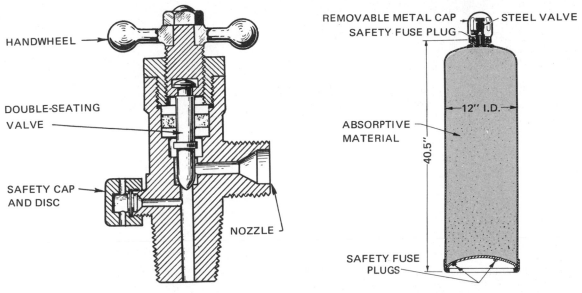

Fig. 2-2 Oxygen cylinder valve   Fig. 2-3 Acetylene cylinder

ACETYLENE CYLINDERS

The acetylene cylinder is a welded steel tube. It is filled with a spongy material such as balsa wood or some other absorptive material which is saturated with a chemical solvent called acetone. Acetone absorbs acetylene gas in much the same manner as water absorbs ammonia gas to produce common household ammonia.

The cylinder is equipped with a valve which can only be opened with a special wrench. Safety regulations require this type of valve on all containers carrying flammable, explosive, or toxic gases. The wrench must be in place whenever the cylinder is in use. Acetylene cylinders are also equipped with a number of *fusible plugs* designed to melt at 220 degrees F. These melt and release the pressure in the event the cylinder is exposed to excessive heat.

Figure 2-3 is a cross section of a common acetylene cylinder. The construction details may vary from one manufacturer to another, but all acetylene cylinders are made to very rigid specifications.

Acetylene cylinders are usually charged to a pressure of 250 pounds per square inch; the large size contains about 275 cubic feet. The steel walls of these cylinders are only .175 inch thick. The precautions set forth for oxygen cylinders should be observed with acetylene cylinders. Escaping acetylene mixed with air forms a highly explosive mixture.

REVIEW QUESTIONS

1. What prevents unauthorized persons from opening acetylene valves?

2. Why must an oxygen cylinder valve be fully opened?

3. What happens to the pressure in a cylinder as the temperature is raised?

   What happens when the temperature is lowered?

4. Suppose the protective cap is left off a fully-charged oxygen cylinder and an accident causes the valve to be broken off:

   a. What is the force of the gas per square inch?

   b. Is this enough force to cause the cylinder to move?

5. What is the function of the cylinder valve?

UNIT 3   WELDING GASES

OXYGEN

Flame is produced by combining oxygen with other materials. When the air we breathe, which is only one-fifth oxygen, combines with other elements to produce a flame, this flame is low in temperature and the rate of burning is rather slow.

However, if pure oxygen is substituted for air, the burning is much more rapid and the temperature is much higher. Oil in the regulators, hoses, torches, or even in open air burns with explosive rapidity when exposed to pure oxygen.

CAUTION: Oxygen must never be allowed to come in contact with any flammable material without proper controls and equipment. The use of oxygen to blow dust and dirt from working surfaces or from a worker's hair or clothing is extremely dangerous.

Most of the oxygen produced commercially in the United States is made by liquefying air and then recovering the pure oxygen. The oxygen thus produced is of such high purity that it can be used not only to produce the most efficient flame for welding and flame cutting operations, but also for medical purposes.

Oxygen is a colorless, odorless, tasteless gas which is slightly heavier than air. The weight of 12.07 cubic feet of oxygen at atmospheric pressure and 70 degrees F. is one pound.

ELECTRIC ARC	19,832	F	11,000	C
SURFACE OF SUN	10,832	F	6,000	C
OXYACETYLENE FLAME	5,900	F	3,260	C
OXYHYDROGEN FLAME	4,752	F	2,900	C
INTERIOR OF INTERNAL COMBUSTION ENGINE	3,272	F	1,800	C
COPPER MELTS	1,976	F	1,080	C
MAGNESIUM MELTS	1,204	F	651	C
WATER BOILS	212	F	100	C
ICE MELTS	32	F	0	C
LIQUID AIR BOILS	-292	F	-180	C
LIQUID HELIUM BOILS	-452	F	-269	C
ABSOLUTE ZERO	-459.4	F	-273	C

Fig. 3-1 Some comparative temperatures

ACETYLENE

Acetylene gas is a chemical compound composed of carbon and hydrogen. It combines with oxygen to produce the hottest gas flame known. Unfortunately, acetylene is an unstable compound and must be handled properly to avoid explosions.

Unstable acetylene gas tends to break down chemically when under a pressure greater than 15 pounds per square inch. This chemical breakdown produces great amounts of heat; the resulting high pressure develops so rapidly that a violent explosion may result.

Acetylene gas which is dissolved in acetone does not tend to break down chemically and can be used with complete safety. However, any attempt to compress acetylene in a free state in hoses, pipes, or cylinders at a pressure greater than 15 pounds per square inch can be very dangerous.

Acetylene is produced by dissolving calcium carbide in water. This process should be carried out only in approved generators. One pound of calcium carbide produces 4.5 cubic feet of acetylene gas. Acetylene is made up of two atoms of carbon and two atoms of hydrogen. It has a distinctive odor. The weight of 14.5 cubic feet is one pound. The amount dissolved in an acetylene cylinder is determined by weighing the cylinder and contents, subtracting the weight of the empty cylinder, and multiplying the remainder, which is the weight of the gas, by 14.5. The empty cylinder weight is always stamped into the cylinder.

REVIEW QUESTIONS

1. Why is it dangerous to place calcium carbide and water in a closed container and generate acetylene gas?

2. What is the probable effect if oil or grease is allowed to come in contact with oxygen in the regulators or cylinders?

3. What element must always be present if a flame is to be produced and maintained?

4. At what pressure will acetylene gas become unstable in a free state?

5. Can oxygen be referred to as air? Why?

UNIT 4   OXYGEN AND ACETYLENE REGULATORS

REGULATORS

Oxygen and acetylene *regulators* reduce the high cylinder pressures, safely and efficiently, to usable working pressures. Regulators also maintain these pressures within very close limits under varying conditions of demand.

Figure 4-1 shows the relatively simple operation of a regulator. The pressure in the hoses is controlled by applying pressure to the spring through an adjusting screw. The spring applies pressure to a flexible rubber diaphragm which is connected to the high-pressure valve. The gas from the cylinder flowing through this valve builds up pressure behind the diaphragm. When this pressure equals the pressure of the spring, the valve closes and shuts off the flow of gas to the diaphragm area. When the pressure in this area is reduced by drawing gas from the regulator to the torch the spring opens the valve again.

Regulators are made to rigid specifications from the finest of materials, and are equipped with safety devices to prevent injury to the operator or the equipment. All regulators are equipped with ball-check safety valves or bursting discs to prevent pressure buildup within the regulator, hoses, or torch.

GAGES

Most regulators are equipped with gages which indicate the amount of pressure in the cylinder and the working pressure in the hoses and torch.

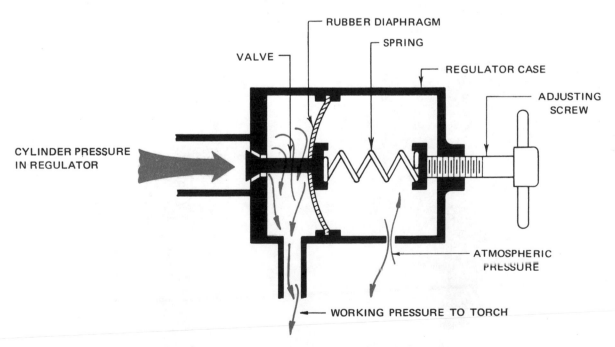

Fig. 4-1 Construction details of a single-stage regulator

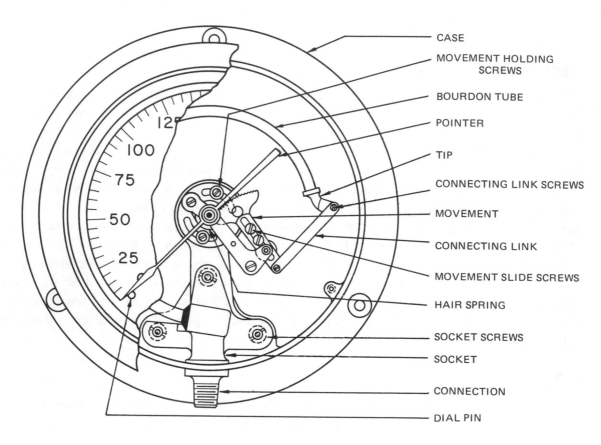

CASE

MOVEMENT HOLDING
SCREWS

BOURDON TUBE

POINTER

TIP

CONNECTING LINK SCREWS

MOVEMENT

CONNECTING LINK

MOVEMENT SLIDE SCREWS

HAIR SPRING

SOCKET SCREWS

SOCKET

CONNECTION

DIAL PIN

Fig. 4-2 Construction details of a gage

Fig. 4-3 Oxygen regulator

Fig. 4-4 Acetylene regulator

These gages have very thin backs which open to release the pressure if the Bourdon tube ruptures. This tube is essential to the operation of each gage. If this precaution were not taken, excessive pressure in the gage case could cause the glass front of the gage to explode and injure the operator.

Since gages frequently get out of calibration, they are only indicators of cylinder and working pressures. Regulators work regardless of the accuracy of the gages.

Unit 8 describes the procedure that should be followed to insure safety and efficiency when adjusting the regulators, regardless of the pressures indicated on the gages.

REVIEW QUESTIONS

1. What two purposes do oxygen and acetylene regulators serve?

2. What purpose do gages on regulators serve?

3. Should a regulator adjusting screw be turned all the way in when the regulator is to be turned off?

4. What safety devices guard against excessive pressure in regulator cases and in hoses?

5. Can a regulator be accurate if the gage is damaged?

UNIT 5   TYPES AND USES OF WELDING TORCHES

The body of the welding torch serves as a handle so the operator can hold and direct the flame. Beyond the handle, the torch is equipped with a means of attaching the mixing head and welding tip, figure 5-1.

The accurately sized holes in welding and cutting tips are called *orifices.* The purpose of the *mixing head* is to combine the two welding gases into a usable form. The only mixed oxygen and acetylene is that amount contained in the space from the mixing head to the tip orifice. In most cases, it represents a very small portion of a cubic inch. This keeps the amount of this highly explosive mixture within safe limits. Any attempt to mix greater amounts may result in violent explosions.

Two types of torches are in common use. In the *injector-type torch,* the acetylene at low pressure is carried through the torch and tip by the force of the higher oxygen pressure through a venturi-type device, shown in figure 5-2. The mixing head and injector are usually a part of the tip which the operator changes according to the size needed.

In the *medium-pressure torch,* figure 5-3, both gases are delivered through the torch to the tip at equal pressures. In this type of torch, the mixer or mixing head is usually a separate piece into which a variety of tips may be fitted.

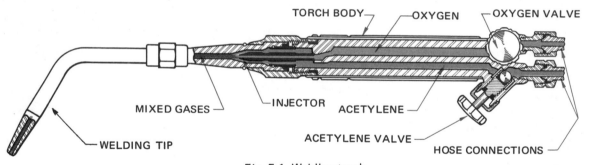

Fig. 5-1 Welding torch

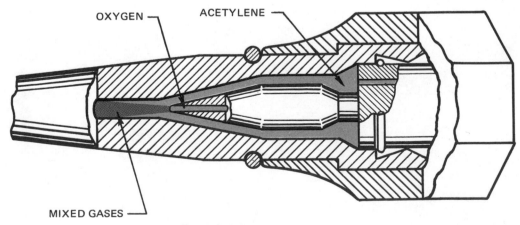

Fig. 5-2 Injector-type mixer

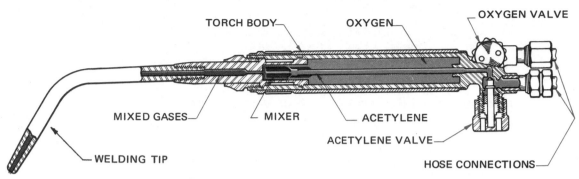

TORCH BODY OXYGEN OXYGEN VALVE

MIXED GASES MIXER ACETYLENE

ACETYLENE VALVE

WELDING TIP HOSE CONNECTIONS

Fig. 5-3 Medium-pressure torch

Fig. 5-4 Welding torch

All types of torches are equipped with a pair of needle valves to turn the welding gases on and off, and to make small pressure adjustments.

REVIEW QUESTIONS

1. What is the chief difference between the two types of torches?

2. a. What is the basic purpose of torch needle valves?

   b. What secondary purpose do they serve?

3. From the construction details indicated in the text, which type of torch is probably more costly?

4. What is the proper name for the holes in welding and cutting tips?

5. How many valves do most torches have?

UNIT 6   WELDING TIPS

The purpose of all welding tips is to provide a safe, convenient method of varying the amount of heat supplied to the weld. They also provide a convenient method of directing the flame and heat to the exact place the operator chooses.

SELECTION OF SIZES

To provide for different amounts of heat, welding tips are made in various sizes. The size is determined by the drill size of the orifice in the tip. As the orifice size increases, greater amounts of the welding gases pass through and are burned to supply a greater amount of heat. However, the temperature of the neutral oxyacetylene flame always remains at 5,900 degrees F., regardless of the quantity of heat provided.

The choice of the proper tip size is very important to good welding. All manufacturers of welding torches supply a chart of recommended sizes for various thicknesses of metal. They also recommend oxygen and acetylene pressures for various types and sizes of tips. These tables provide orifice sizes and the proper drill size for cleaning each orifice, figure 6-1.

CARE OF TIPS

All welding tips are made of copper and may be damaged by careless handling. Dropping, prying, or hammering, the tips on the work may make them unfit for further use. It is im-

PLATE THICKNESSES		ORIFICE SIZE	GAS PRESSURES IN P.S.I.			
			INJECTOR-TYPE TORCH		EQUAL-PRESSURE TORCH	
GAGE	INCHES	NO. DRILL	ACETYLENE	OXYGEN	ACETYLENE	OXYGEN
32	.010	74	5	5 – 7	1	1
28	.016	70	5	7 – 8	1	1
26	.019	70	5	7 – 10	1	1
22	$\frac{1}{32}$	65	5	7 – 18	2	2
16	$\frac{1}{16}$	56	5	8 – 20	3	3
13	$\frac{3}{32}$	56 – 54	5	15 – 20	4	4
11	$\frac{1}{8}$	54 – 53	5	12 – 24	4	4
8	$\frac{3}{16}$	53 – 50	5	16 – 25	5	5
	$\frac{1}{4}$	50 – 46	5	20 – 29	6	6
	$\frac{3}{8}$	46 – 44	5	24 – 33	7	7
	$\frac{1}{2}$	40	5	29 – 34	8	8
	$\frac{5}{8}$	30	5	30 – 40	9	9
	$\frac{3}{4}$	30 – 29	5	30 – 40	10	10
	1	23	5	30 – 42	12	12

Fig. 6-1 Relation of plate thickness, orifice size, and gas pressure

DRILL SIZE NO.	AIRCO – ALL	CRAFTSMAN – ALL	DOCKSON – 4EC, 4SC	7EC	GASWELD – G25, G35	G55	AVG	HARRIS – 13,14,16,17,50	OTHERS	K-G – EUS, KUS, KS	MARQUETTE – A, AL	B, BL	MECO – ALL	MILBURN – W-200	W-11	W-600	NATIONAL – G	P	R	OXWELD – W-29	W-17	PUROX – 33	34	35	REGO – ALL	SMITH – LIFETIME	NO. 5	NO. 2	VICTOR – ALL
80													00	00		0000													
79					00	00											00					0							
78			1	1					00			00B		0														18	
77					0	0				0						000													
76										0			0	1			0	0										19	
75	00							75			00	OB								000			1						000
74					00	1	1		00					00									1					A20	
73			2	2																				1					000½
72	0	0			0			72				1B						1	1						72				
71									0					000												B60	50	A21	
70		1							0				1							00	2								00
69				1	2	2							2			0													
68	1							68	1			2B		00									2		68			A22	
67																													00½
66			3	3					1		1						2	2					2	2					
65		2				3	2		2				2			1				0	4	4						A23	0
64																													
63	2								2					0												B61	51		0½
62		2						62	2			3B	3			2									62				
61																													
60							4														1		3						1
59																													
58			4	4	3	3		58	3		3					3									58	B62	52	A25	
57							5										3	4	1					3					1½
56	3							56	4			4B								2	6	6	4					A26	2
55			5	5	4	4			4					4			5			4	3	3			55				2½
54		3			5	5	6				4	5B				2				3	9	9		4		B63	53	A27	
53	4						5	53	5				5			5				12	12	5		5	53				3
52			6	6	6	6	7						5													B64	54	A28	
51	5								6			6B		6	3			4	4										3½
50							8	50	5							6				15	15	6	6		50			A29	
49																													4
48	6	4	7	7	7				7			7B		7	4						20					B65	55	A210	
47									8		6		6							4			7						
46																	5	5		20					46				
45								45	8																				
44	7		8	8	8	8								5	7						30		8			B66	56		
43									9	9	7			8						5			8						5
42		5			9						7										30				42				
41					9									6	8			6	6										
40	8				10			10	10	40	8			9							40				40	B67	57		
39																			7					9					
38					11										7														
37		6																											
36			10		12						9		8	10	8				8				10		36	B68			6
35	9				13			35							9	9													
34																													
33		7			14										9				6										
32			11						12		10							9								B69			
31					15									11		10							11	31					
30	10		12		16			30	15		11			12	10				10	7		55	13						7
29		8			16								9	13						8	70								8
28			13								10	12							9						B610				9
27									19		12								10										10
26			14																11						B611				11
25	11							25						14	13				12						25				12
24			15						22			14														B612			
23																					90		15						
22													11	14															
21															15														
20	12							20						15											20				
19														16															
18													12																

NOTE: USE A DRILL ONE SIZE SMALLER FOR CLEANING ORIFICES

Fig. 6-2 Comparison guide for welding tip sizes

portant to clean the tip orifice with the proper tip drill. The use of an incorrect drill or procedure can ruin a tip.

REVIEW QUESTIONS

Note: Determine the proper orifice size from figure 6-1. Then find the tip number
closest to this size in figure 6-2.

1. What size Craftsman® tip should be used for welding 22-gage steel?

2. What size Rego® tip should be used for welding 1/8-inch plate?

3. What size Victor® tip should be used for welding 16-gage steel?

4. What size Airco® tip should be used for welding 1/4-inch steel?

5. What size Oxweld® W-17 tip should be used for welding 1/2-inch steel?

6. What size Smith-Lifetime® tip should be used for welding 8-gage steel?

UNIT 7   THE OXYACETYLENE WELDING FLAME

The flame is the actual tool of oxyacetylene welding. All of the welding equipment merely serves to maintain and control the flame.

The flame must be of the proper size, shape, and condition in order to operate with maximum efficiency. The oxyacetylene flame differs from most other types of tools in that it is not ready-made. The operator must produce the proper flame each time he lights the torch.

Once the operator masters the adjustment of the flame, his ability as a welder increases in direct proportion to the amount of practice he has.

TYPES OF FLAMES

The oxyacetylene flame can be adjusted to produce three distinctly different types of flame. Each of these types has a very marked effect on the metal being fused or welded. In the order of their general use, the flames are *neutral, carburizing,* and *oxidizing*. Figure 7-1 illustrates their shapes and characteristics.

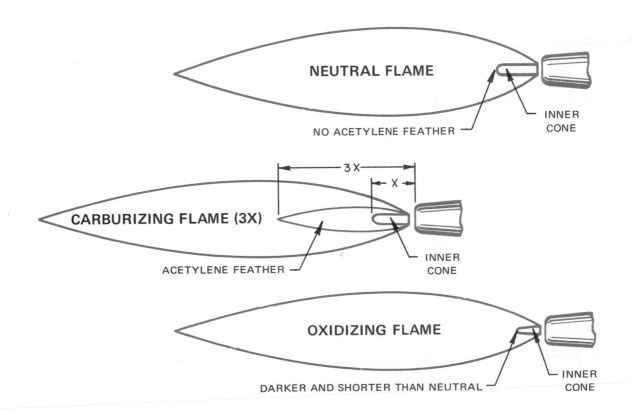

Fig. 7-1 Types of flames

The *neutral* flame is one in which equal amounts of oxygen and acetylene combine in the inner cone to produce a flame with a temperature of 5,900 degrees F. The inner cone is light blue in color. It is surrounded by an outer flame envelope, produced by the combination of oxygen in the air and superheated carbon monoxide and hydrogen gases from the inner cone. This envelope is usually a much darker blue than the inner cone. The advantage of the neutral flame is that it adds nothing to the metal and takes nothing away. Once the metal has been fused, it is chemically the same as before welding.

The *carburizing* flame is indicated by streamers of excess acetylene from the inner cone. These streamers are usually called *feathers* of acetylene, or simply the *acetylene feather.* The feather length depends on the amount of excess acetylene. The outer flame envelope is longer than that of the neutral flame and is usually much brighter in color. This excess acetylene is very rich in carbon. When carbon is applied to red-hot or molten metal, it tends to combine with steel and iron to produce the very hard, brittle substance known as iron carbide. This chemical change leaves the metal in the weld unfit for many applications in which the weld may need to be bent or stretched. While this type of flame does have its uses, it should be avoided when fusion welding those metals which tend to absorb carbon.

The carburizing flame in figure 7-1 shows the relation of the acetylene streamers to the inner cone. Job conditions sometime require an excess of acetylene in terms of the length of the inner cone. Figure 7-1.

The *oxidizing* flame, which has an excess of oxygen, is probably the least used of any of the three flames. In appearance, the inner cone is shorter, much bluer in color, and usually more pointed than a neutral flame. The outer flame envelope is much shorter and tends to fan out at the end. The neutral and carburizing envelopes tend to come to a sharp point. The excess oxygen in the flame causes the temperature to rise as high as 6,300 degrees F. This temperature would be an advantage if it were not for the fact that the excess oxygen, expecially at high temperatures, tends to combine with many metals to form hard, brittle, low-strength oxides. For this reason, even slightly oxidizing flames should be avoided in welding.

REVIEW QUESTIONS

1. What chemical change takes place when a carburizing flame is used to weld steel?

2. What substance is produced when an oxidizing flame is used to weld steel?

3. Make a labeled and dimensioned sketch of a 2X flame. What are the significant parts?

4. Make a labeled and dimensioned sketch of a 3 1/2X flame. What are the significant parts?

5. What is the temperature of a neutral flame?

UNIT 8 SETTING UP EQUIPMENT AND LIGHTING THE TORCH

Oxyacetylene welding equipment must be set up frequently and it must be done efficiently. Since hazards are present, each step must be performed correctly. The proper sequence must be followed to insure maximum safety to personnel and equipment.

The cylinder caps are removed and put in their proper place. The cylinders should be fastened to a wall or other structure with chains, straps or bars, to prevent them from being tipped over. To use oxygen and acetylene cylinders and equipment without this safety precaution is to invite damage to the equipment and injury to the operator.

PROCEDURE

1. Aim the cylinder nozzle so it does not blow toward anyone. Crack the valve on each cylinder by opening the valve and closing it quickly. This blows any dust or other foreign material from the nozzle.

2. Attach the regulators to the cylinder nozzles.

   Note: All oxygen regulators in commercial use have a standard *right-hand* thread and fit all standard oxygen cylinders. Acetylene regulators may have *right- or left-hand* threads and may have either a male or female connection, depending on the company supplying the gas. Adapters of various types may be needed to fit the existing regulators to different acetylene cylinders.

3. Attach the hoses to the regulators.

   Note: All oxygen hose connections have *right-hand* threads. All acetylene hose connections have *left-hand* threads. The acetylene hose connection nuts are distinguished from the oxygen nuts by a *groove* machined around the center of the nut figure 8-1.

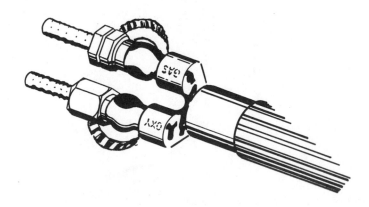

Fig. 8-1 Oxygen and acetylene hose connections

4. Attach the torch to the other end of the hoses noting that while the hose connections may be a different size at the torch than at the regulators, they still have right- and left-hand threads.

    Note: Use only the wrenches provided for attaching hoses and regulators. These wrenches are designed to give the proper leverage to tighten the joints without putting undue strain on the equipment. If the joints cannot be properly tightened, something is wrong.

5. Select the proper tip and mixing head and attach it to the torch. Position the tip so that the needle valves are on the side or bottom of the torch when the tip is in the proper welding position.

6. Back off the regulator screws on both units until the screws turn freely. This is necessary to eliminate a sudden surge of excessive pressure on the working side of the regulator when the cylinder is turned on.

7. Be sure both torch needle valves are turned off (clockwise). This is an added safety precaution to make sure excessive pressure cannot be backed through the mixing head and into the opposite hose.

8. Open the acetylene cylinder valve 1/4 to 1/2 turn. Open the oxygen cylinder valve all the way.

9. Open the acetylene needle valve one full turn. Turn the adjusting screw on the acetylene regulator clockwise until gas comes from the tip. Light this gas with a sparklighter.

10. Adjust the regulator screw until there is a gap of about 1/4 inch between the tip and the flame. This is the proper pressure for the size of tip being used regardless of the gage pressure shown on the working pressure gage.

11. Open the oxygen needle valve on the torch one full turn. Turn the oxygen regulator adjusting screw clockwise until the flame changes appearance as oxygen is mixed with the acetylene.

12. Continue to turn the adjusting screw until the feather of acetylene just disappears into the end of the inner cone. This produces a neutral flame which is used in most welding.

This procedure for adjusting the oxyacetylene flame is the safest method of insuring the proper working pressures in both hoses and tip. Working pressure gages are delicate and easily get out of calibration. If this happens, excessive pressure can be built up in the hoses before it is discovered. However, if the pressures are adapted to the flame as indicated, there are equal pressures in both hoses which eliminates the possibility of backing gas from one hose to the other to form an explosive mixture. With the regulators properly adjusted, minor flame adjustments are made with the torch needle valves.

When the welding or cutting operation is finished, close the torch acetylene valve first, then the torch oxygen valve.

To shut down the equipment for an extended period of time, such as overnight, it should be purged. Use the following procedure:

1. Close the oxygen cylinder valve.
2. Open the torch oxygen valve to release all pressure from the hose and regulator.

3. Turn out the pressure adjusting screw of the oxygen regulator.
4. Close the torch oxygen valve.
5. Follow the same sequence for purging acetylene.

REVIEW QUESTIONS

1. What are the steps necessary to turn on the welding gases properly and safely and adjust them to a suitable flame?

2. What steps are necessary to assemble an oxyacetylene outfit for welding?

3. Why should the needle valves on the torch be turned off at a particular step in the sequence rather than at some other time?

4. Could excessive oxygen pressure backed through the torch cause an explosion in the acetylene hose without outside ignition? Explain. (Refer to Unit 3 — Acetylene Gas.)

5. How is a left-hand nut different from a right-hand nut in appearance?

UNIT 9   FLAME CUTTING

One of the fastest ways of cutting ferrous metals is by the use of the oxyacetylene torch. Other advantages of this cutting method are:

1. A relatively smooth cut is produced.
2. Very thick steel (over 4 feet) can be cut.
3. The equipment is portable.
4. Underwater cutting is possible with some adaptations.
5. The equipment lends itself to automatic processes in manufacturing.

The terms "cutting" and "burning" are used interchangeably to describe this process.

THE BURNING PROCESS

Oxyacetylene flame cutting is actually a burning process in which the metal to be cut is heated on the surface to the kindling temperature of steel, (1,600 - 1,800 degrees F.). A small stream of pure oxygen is then directed at the work. The oxygen causes the metal to ignite and burn to produce more heat. This additional heat causes the nearby metal to burn so that the process is continuous once it has started.

Only those ferrous metals which oxidize rapidly can be flame-cut. These metals include all the straight carbon steels and many of the alloys. Stainless steels and most of the so-called high-speed steels cannot be flame-cut.

EQUIPMENT

Cutting is done with a special torch fitted with interchangeable tips so that it can be adapted to cut a wide variety of metal thicknesses. The torch and tips are constructed so that they can preheat the work to the kindling temperature. The torch also includes a lever for turning on and stopping the stream of high-pressure cutting oxygen as required. The torch is usually made of forged brass and brass tubing, figure 9-1.

For hand-manipulated flame-cutting operations, the tips are made of copper. If the tip of the torch is used as a hammer, lever, or crowbar, permanent damage is done.

COMPARISON CHARTS

Because cutting torch tips are interchangeable, chart 9-1 may be used for the torch tips of all major manufacturers.

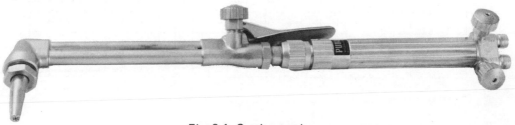

Fig. 9-1 Cutting torch

Drill Size No. for Oxygen Cutting Orifice	AIRCO—ALL EXCEPT #45	CRAFTSMAN—B	DOCKSON—ALL	GASWELD—ALL	HC-31,32	HC-39	WC-20, 35	WC-10, 55	HARRIS 2890-F	6290	7490-A	K-G, M4, M5	MARQUETTE E	C, D, DI, 4 PREHEAT	C, D, DI, 6 PREHEAT	MECO—ALL	MILBURN X100	X 2000	X 2300	NATIONAL—ALL	OXWELD CW-29	CW-23, C-31, 32	PUROX-33	34, 35	REGO—AW	SMITH LIFETIME 4 PH	LIFETIME 6 PREHEAT	VICTOR—ALL
76																				2	2							
75																												
74																												
73																												000
72																												
71									00																			
70																												
69			00	0				00																				
68	00										68										3			68				
67																0												
66			0																				0					
65									0			1B																00
64																												0
63																	00	00										
62	0								1	00	62												1	62				
61		1																										
60	1											2B		0A	0A	0	0			0	4	4	1			0		
59				2																								0
58																												0
57																1		1	1	1								
56	1					1		1	1	1	56		1A	1A										56		1	1	
55		2	2												1													
54	2													2		2	2	2					2			2		1
53	3										53										6			53				
52	3		2				2							2A		2										3	2	2
51								2																51				
50											50						3	3	3				3					
49	4	4	3						3				2A															3
48																3	4									4	3	
47																												
46														3						4	8			46				
45	5	5	3				3	3			45					4	5	4										4
44																									4			
43			4											3A														
42								4								5		5						42				
41	6																5											
40	6										40					6	6										4	5
39																				10								
38													3A															
37														4A		7	7			6								
36		5	4												4		6											
35							4	5			35					8	8							35				6
34	7	7		5				4																				
33																												
32		6						5								9		7										
31																				12								
30	8						5	6			30		4A	5A	5	10		7	8					30		5		
29																	11											
28		7														6	12											7
27		8																										
26	9													6A				8										
25											25													25				
24																											6	
23																												
22																		9										
21																												
20																												
19																												
18	10																10											
17																	11											

NOTE: USE A DRILL ONE SIZE SMALLER FOR CLEANING ORIFICES.

Chart 9-1 Comparison guide of cutting tip sizes

THICKNESS OF STEEL	1/4"	3/8"	1/2"	3/4"	1"	1/4"	1 1/2"	2"	2 1/2"	3"	4"	5"	6"
AIRCO TIP SIZE	0	1	1	2	2	2	3	3	4	5	5	6	6
GAGE PRESSURE OXYGEN P.S.I.	30	30	40	40	50	60	45	50	50	45	60	50	55
GAGE PRESSURE ACETYLENE P.S.I.	3	3	3	3	3	3	3	3	3	4	4	5	5
SPEED IN INCHES PER MIN.	20	19	17	15	14	13	12	10	9	8	7	6	5
OXYGEN CONSUMPTION CU. FT. PER HOUR	50	75	90	120	140	160	185	200	250	310	385	460	495
ACETYLENE CONSUMPTION CU. FT. PER HOUR	9	12	12	14	14	14	16	16	17	22	22	28	28
APPROXIMATE WIDTH OF KERF IN INCHES	.075	.095	.095	.110	.110	.110	.130	.130	.145	.165	.165	.190	.190
CUTTING ORIFICE CLEANING DRILL SIZE	64	57	57	55	55	55	53	53	50	47	47	42	42
PREHEAT ORIFICE CLEANING DRILL	71	69	69	68	68	68	66	66	65	63	63	61	61

NOTE: This chart pertains to Airco Style 124 tips only. If equipment from other manufacturers is used, refer to the chart, "Comparison Guide of Cutting Tip Sizes" and choose a tip with a cutting or orifice size comparable to that indicated above.

Chart 9-2 Relation of cutting tip size to plate thickness

Chart 9-2, refers to Airco® style 124 tips only. This chart gives proper tip sizes, cutting oxygen pressures, and rate of travel for the various thicknesses of metal. If equipment from other manufacturers is used, refer first to the Comparison Chart and choose a tip with a cutting orifice size close to the Airco® size.

HAZARDS

The operator must protect his eyes at all times with goggles fitted with proper lenses, usually shade 5 or 6. Gauntlet-type gloves and any other equipment necessary to give protection from the molten iron oxide, should be worn.

Since the high-pressure stream of cutting oxygen can throw small bits of molten oxide, at a temperature of 3,000 degrees F. and up, for distances of 50-60 feet, the operator should check before starting the burning operation to be sure that all flammable and explosive materials have been removed to a safe place.

The operator should insure that all personnel in the area are warned of the shower of molten metal that will occur so that they may take the necessary precautions.

The International Acetylene Association and the Underwriters' Laboratories recommend that an additional workman with fire-fighting equipment be assigned to each unit during cutting and for 2 hours after completion of cutting to guard against fires.

REVIEW QUESTIONS

1. What are the limitations of flame cutting?

2. What chemical change takes place during the burning process?

3. What special safety precautions must be taken?

4. How wide must a piece of one-inch steel plate be so that 10 strips each 3 inches wide can be cut from it? Make the proper allowance for the width of the kerf (cut width) from chart 9-2.

5. Using chart 9-2 determine how much time is required to cut the plates in question 4 if they are each 10 feet long? Figure actual cutting time only.

6. How much oxygen and acetylene are used in problems 4 and 5? Figure only the actual time required to make the cuts.

UNIT 10   STRAIGHT LINE CUTTING

Several things affect the speed, smoothness, and general quality of a cut made by an oxyacetylene flame. This unit provides practice in changing these variables to determine the best methods for flame cutting.

The actual cutting process demonstrates the danger of personal burns and fires which might cause property damage.

MATERIALS

1/4-inch or 3/8-inch thick steel plate, approximately 4 in. x 10 in.

Cutting torch fitted with an Airco® #0 or #1 cutting tip or comparable equipment. See chart 1, Comparison of Cutting Tip Sizes, page 24.

PROCEDURE

1. Draw a series of straight parallel lines about 2 inches apart on the plate. Use soapstone for marking so that the lines show up when the cutting goggles are being worn.

2. Light and adjust the preheating flame to neutral using the data supplied in chart 9-2, "Relation of Cutting Tip Size to Plate Thickness."

3. Start the cut by holding the tip over the edge of the metal so that the vertical centerline of the tip is square with the work and in line with the edge of the plate. The tip is positioned in the torch as indicated in figure 10-1.

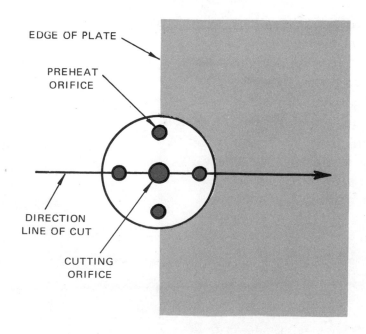

EDGE OF PLATE

PREHEAT ORIFICE

DIRECTION LINE OF CUT

CUTTING ORIFICE

Fig. 10-1 Tip alignment for square cuts

1. This is a correctly made cut in 1-in. plate. The edge is square and the draglines are vertical and not too pronounced.
2. Preheat flames were too small for this cut with the result that the cutting speed was too slow, causing gouging at the bottom.
3. Preheat flames were too long with the result that the top surface has melted over, the cut edge is rough, and there is an excessive amount of adhering slag.
4. Oxygen pressure was too low with the result that the top edge has melted over because of the too slow cutting speed.
5. Oxygen pressure was too high and the nozzle size too small with the result that the entire control of the cut has been lost.
6. Cutting speed was too slow with the result that the irregularities of the draglines are emphasized.
7. Cutting speed was too high with the result that there is a pronounced break to the dragline and the cut edge is rough.
8. Torch travel was unsteady with the result that the cut edge is wavy and rough.
9. Cut was lost and not carefully restarted with the result that bad gouges were caused at the restarting point.
10. Correct procedure was used in making this cut.
11. Too much preheat was used and the nozzle was held too close to the plate with the result that a bad melting over of the top edge occured.
12. Too little preheat was used and the flames were held too far from the plate with the result that the heat spread opened up the kerf at the top. The kerf is too wide at the top and tapers in.

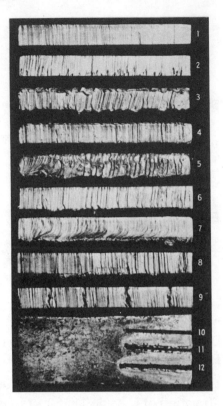

Fig. 10-2 Common faults that occur in hand cutting

4. When the edge of the work becomes bright red, turn the cutting oxygen on with the lever. Note that the oxygen makes a cut through the plate at the same angle that the centerline of the tip makes with the work.

5. Continue the cut, making sure that the tip is square with the work. Observe that when the rate of travel is right, the slag or iron oxide coming from the cut makes a sound like cloth being torn. The tearing sound serves as a guide to the correct rate of travel in most manual flame-cutting operations.

6. Finish the cut and check the flame-cut edge for smoothness, straightness, and amount of slag on the bottom edge of the cut surfaces.

7. Make more cuts but vary the amount of preheating by decreasing and increasing the acetylene pressure before each cut. Observe the finished cut for smoothness, melting of the top edge of the plate, amount of slag on the bottom edge of the plate, and ease of removal of this slag. Compare the plates cut and determine which amount of preheat produces the best results.

8. Make more cuts but vary the rate of travel from very slow to normal to very fast. Ob-

Fig. 10-3 Straight cut

serve these finished cuts and check the appearance of the top and bottom of each plate, and also the ease of slag removal. Determine which rate of travel produces the best results.

9. Make more cuts but vary the amount of cutting oxygen pressure from low to normal to high and check the results as in step 8.

10. Make one or two cuts with the tip perpendicular to the work but move the torch so that the tip zigzags along the straight line drawn on the plate. Notice that the surface of the cut edge follows the amount and direction the tip moves from the straight line.

REVIEW QUESTIONS

1. A number of variables have been tried out in this unit. What conclusions can be drawn about the importance of each of these variables with regard to the ability to make straight line cuts?

   a. Tip angle

   b. Amount of preheat

   c. Amount of cutting oxygen pressure

   d. Rate of travel

   e. Direction of travel

UNIT 11   BEVEL CUTTING

Making bevel cuts on steel plate is a common cutting operation. The technique is similar to that used for making straight cuts.

Skill in cutting operations is gained only through practice and with a definite goal in mind. Unguided wanderings over a plate add very little to an operator's skill and waste material and gas.

MATERIALS

3/8-inch thick steel plate
Airco® cutting torch fitted with a #1 tip or comparable equipment

PROCEDURE

1. Draw a series of parallel straight lines spaced on 2-inch centers on the work with soapstone.

   Note: These lines are sometimes centerpunched at close intervals to improve visibility. This improves accuracy, but usually results in a slight loss of quality in the cut. The center punch marks cause the high-speed, high-pressure cutting oxygen to stray somewhat.

2. Hold the cutting tip at an angle of 45 degrees with the work and keep this angle when bevel cutting.

3. Proceed with the cut in the same manner as in unit 10. The cut progresses better if the preheating orifices are aligned as in figure 11-1.

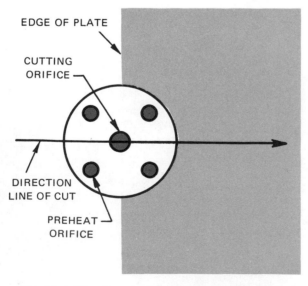

Fig. 11-1 Tip alignment for bevel cuts

Fig. 11-2 Bevel cut

4. Inspect the finished cut for smoothness and uniformity of angle. Check the amount and ease of removal of slag.

5. Make more bevel cuts, but correct the variables until good results are obtained each time.

6. Make another cut, but as the tip moves forward, bring it alternately closer and farther from the work and observe the results.

REVIEW QUESTIONS

1. What variable is responsible for the slag being hard to remove from the bottom of the plate?

2. How does the distance of the tip from the work during the burning process affect the appearance of the finished cut?

3. In making the cut on the plate in this unit, how is the proper tip size, gas pressure, and rate of travel determined?

4. Why is a line to be cut center punched?

5. Why is bevel cutting done?

UNIT 12 PIERCING AND HOLE CUTTING

Holes are easily cut in steel plate by the oxyacetylene flame-cutting procedure. This is a fast operation, adaptable to plates of varying thicknesses and holes of varying sizes. It is useful in cutting irregular shapes.

It is recommended that a diamond point chisel be used to turn up a burr at the point at which the cut is to be started. This burr reaches the kindling temperature much faster than the surface of a flat plate. If a large number of holes are to be pierced, this procedure saves large amounts of preheating gas and operator time.

CAUTION: If particular care is not used in this operation, molten metal may be blown in the face of the operator or into the tip of the torch.

MATERIALS

3/8-inch thick steel plate
Airco® #1 cutting tip or comparable equipment

PROCEDURE

1. Pierce the plate, using the sequence of operations shown in figure 12-1.

   a. Hold the tip about 1/4 inch from the work until the surface reaches the kindling temperature, figure 12-1A.

   b. Open the cutting oxygen valve slowly and, as the burning starts, back the tip away from the work to a distance of about 5/8 inch. The tip must be tilted slightly so that the oxide blows away from the operator and does not blow directly back at the tip, figure 12-1B and C.

   c. Hold the tip in this position until a small hole is pierced through the plate, figure 12-1D.

   d. Lower the tip to the normal burning distance and be sure that it is exactly square with the plate. Then move the torch to enlarge the hole, figure 12-1E.

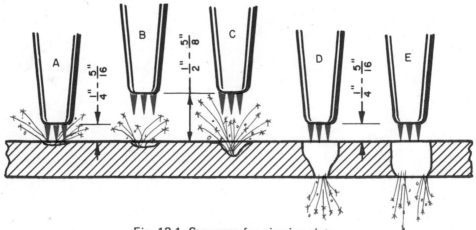

Fig. 12-1 Sequence for piercing plate

2. Continue to move the tip until the hole is the desired size.

   Note: This piercing procedure is recommended for two reasons:

   1. The tip is not ruined by burning the end. This occurs when the tip is held too close to the work for a long period.
   2. The possibility of the molten slag blowing back and clogging the cutting and preheating orifices is reduced.

3. After piercing the plate, move the tip in a circular path to cut a hole of the desired diameter. Considerable practice is necessary to become skillful in making holes which have straight sides, are reasonably round, and close enough to the given diameter to be acceptable.

4. With a pair of dividers, lay out some holes from 1/2 inch to 1 inch in diameter. Center punch the layout line at close intervals to serve as a guide for the cutting operation.

5. Cut the holes, but remember that when cutting to a line the plate should be pierced some distance inside the line. The hole can then be enlarged to the line and the operation completed.

6. Lay out and cut some holes 2 inches and 3 inches in diameter.

   Note: When making large diameter holes, pierce the plate and then move the torch in a straight line until the cut reaches the layout line. Then proceed with the circular cut.

7. Make some round discs. In this case, the piercing operation is performed at a distance outside the layout line. If these discs are to be turned to a specified diameter after cutting, enough material for this operation must be provided.

REVIEW QUESTIONS

1. If a 3-inch round shaft is to be cut off, what preparation is necessary to insure a quick start of the cutting action?

2. How does the procedure vary from the above if the shaft to be cut is square instead of round?

3. If it is desirable to save both the disc and hole when cutting large diameter holes, what procedure should be followed?

4. Why is the tip tilted slightly when starting the cut?

5. Can hole piercing be dangerous for the operator?

UNIT 13 WELDING SYMBOLS

DESCRIBING WELDS ON DRAWINGS

Welding symbols form a shorthand for the draftsman, fabricators, and welding operators. A few good symbols give more information than several paragraphs.

The American Welding Society has prepared a pamphlet, "Symbols for Welding and Nondestructive Testing" (AWS A2.4-76), which indicates to the draftsman the exact procedures and standards to be followed so the fabricators and welding operators may read and understand all the information necessary to produce the correct weld.

The standard AWS symbols for arc and gas welding are shown in figure 13-1.

EXAMPLES OF THE USE OF SYMBOLS

Each of the symbols in this unit should be studied and compared with the drawing which shows its significance. They should also be compared with the symbols shown in figure 13-1.

Throughout this book a symbol related to the particular job is shown together with its meaning. A study of each of these examples will clarify the meaning of the welding symbols.

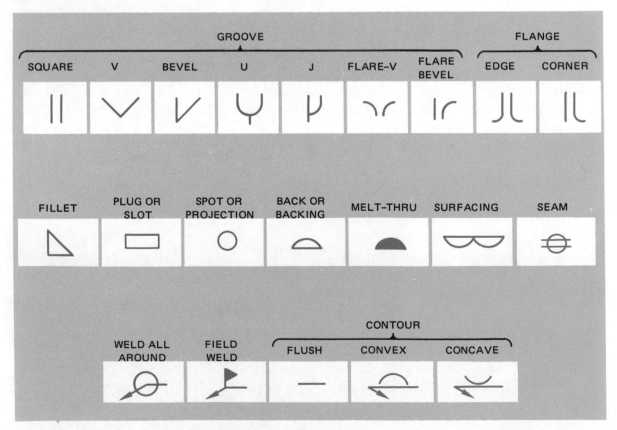

Fig. 13-1 Standard welding symbols

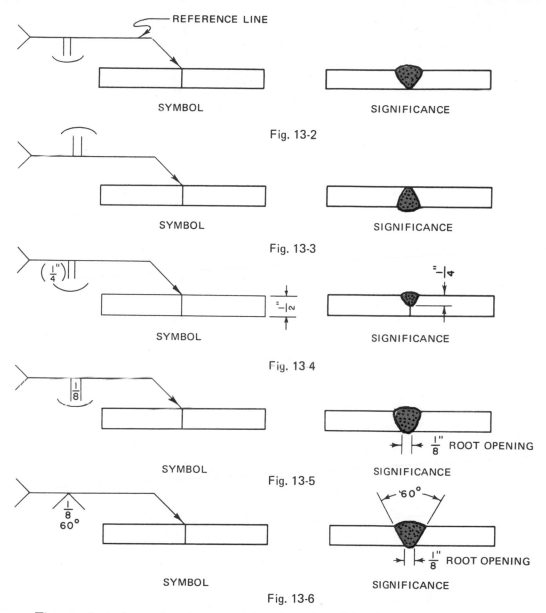

REFERENCE LINE

SYMBOL   SIGNIFICANCE

Fig. 13-2

SYMBOL   SIGNIFICANCE

Fig. 13-3

SYMBOL   SIGNIFICANCE

Fig. 13 4

SYMBOL   SIGNIFICANCE

Fig. 13-5

SYMBOL   SIGNIFICANCE

Fig. 13-6

The symbols from the chart are placed at the mid-point of a reference line. When the symbol is on the near side of the reference line the weld should be made on the arrow side of the joint as in figure 13-2.

If the symbol is on the other side of the reference line, as in figure 13-3, the weld should be made on the far side of the joint or the side opposite the arrowhead.

All penetration and fusion is to be complete unless otherwise indicated by a dimension positioned as shown by the (1/4) in figure 13-4.

To distinguish between root opening and depth of penetration, the amount of root opening for an open square butt joint is indicated by placing the dimension within the symbol, figure 13-5, instead of within parentheses as in the preceding drawing.

The included angle of beveled joints and the root opening is indicated in figure 13-6. If no root opening is indicated on the symbol, it is assumed that the plates are butted tight, unless the manufacturer has set up a standard for all butt joints.

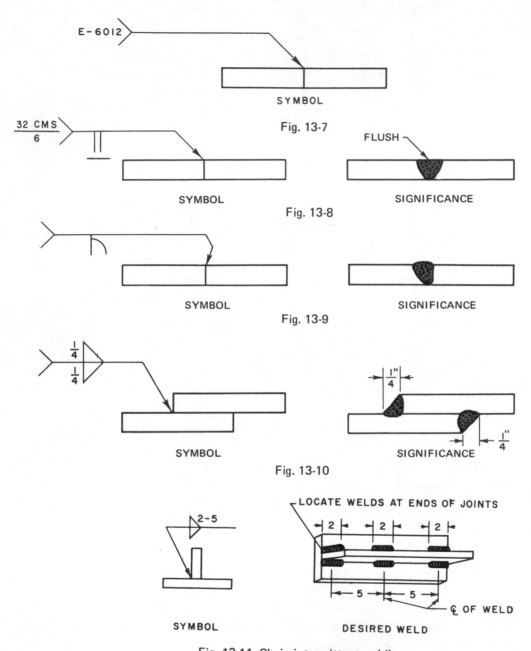

E-6012

SYMBOL

Fig. 13-7

32 CMS / 6

SYMBOL

FLUSH

SIGNIFICANCE

Fig. 13-8

SYMBOL

SIGNIFICANCE

Fig. 13-9

$\frac{1}{4}$
$\frac{1}{4}$

SYMBOL

SIGNIFICANCE

Fig. 13-10

2-5

SYMBOL

LOCATE WELDS AT ENDS OF JOINTS

¢ OF WELD

DESIRED WELD

Fig. 13-11 Chain intermittent welding

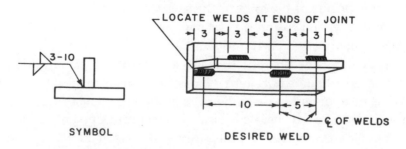

3-10

SYMBOL

LOCATE WELDS AT ENDS OF JOINT

¢ OF WELDS

DESIRED WELD

Fig. 13-12 Staggered intermittent welding

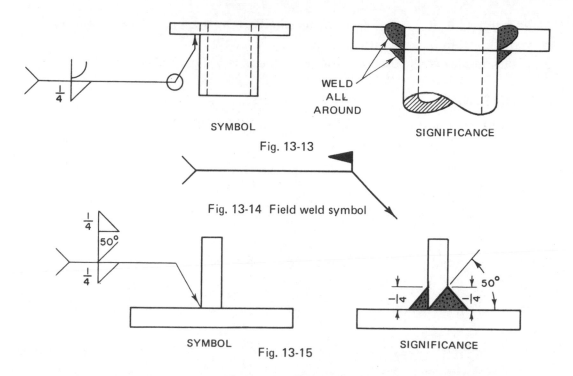

SYMBOL

SIGNIFICANCE

WELD ALL AROUND

Fig. 13-13

Fig. 13-14 Field weld symbol

SYMBOL

SIGNIFICANCE

Fig. 13-15

The tail of the arrow on reference lines is often provided so that a draftsman may indicate a particular specification not otherwise shown by the symbol. Such specifications are usually prepared by individual manufacturers in booklet or loose-leaf form for their engineering and fabricating departments. These specifications cover such items as the welding process to be used (i.e. arc or gas), the size and type of rod or electrode, and the preparation for welding, such as preheating.

Many manufacturers are using the AWS publication, "Symbols for Welding and Nondestructive Testing" which gives very complete rules and examples for welding symbols as well as a complete set of specifications with letters and numbers to indicate the process.

One method of indicating the type of rod to be used is shown in figure 13-7. This shows that the butt weld is to be made with an AWS classification E-6012 electrode.

In figure 13-8, the rod to be used is indicated as a number 32 CMS (carbon mild steel) type; the 6 indicates the size of the rod in 32nds of an inch. In this case it is a 3/16-inch diameter rod. In addition, the symbol indicates that the finished weld is to be flat or flush with the surface of the base metal. This may be accomplished by: G = Grinding, C = Chipping, or M = Milling.

When only one member of a joint is to be beveled, the arrow makes a definite break back toward the member to be beveled, figure 13-9.

The size of fillet and lap beads is indicated in figure 13-10. In all lap and fillet welds, the two legs of the weld are equal unless otherwise specified.

If the welds are to be chain intermittent, the length of the welds and the center-to-center spacing is indicated, as in figure 13-11.

When the weld is to be staggered, the symbol and desired weld is made as in figure 13-12.

An indication that the joint is to be welded all around is shown by using the weld all around symbol at the break in the reference line, as in figure 13-13.

Field welds (any weld not made in the shop) are indicated by placing the field weld symbol at the break in the reference line, as in figure 13-14.

Several symbols may be used together when necessary, figure 13-15.

REVIEW QUESTIONS

1. What is the symbol for a 60-degree closed butt weld on pipe?

2. What is the symbol for a U-groove weld with a 3/32-inch root opening?

3. What is the symbol for a double V, closed butt joint in plate?

4. What is the symbol for a 1/2-inch fillet weld in which a column base is welded to an H-beam all around?

5. What is the symbol for a J-groove weld on the opposite side of a plate joint?

UNIT 14   RUNNING BEADS AND OBSERVING EFFECTS

The quality of the finished weld depends to a large extent on the correct adjustment and use of the flame. This unit provides an opportunity to weld with different kinds of flames and to compare the results. At the same time some acutal welding skill is acquired.

MATERIALS

16- or 18-gage mild steel, 2 to 4 in. wide X 6 to 9 in. long
Airco® #2 welding tip or equivalent

PROCEDURE

1. Light the torch and adjust the flame to neutral.

2. Hold the tip of the inner cone of the flame about 1/8 inch above the work and pointed in the exact direction in which the weld is to proceed. The centerline of the flame should make an angle of 45 to 60 degrees with the work, figure 14-2.

3. Hold the flame in one spot until a puddle of metal 1/4 inch to 3/8 inch in diameter is formed.

4. Proceed with the weld, advancing the flame at a uniform speed in order to keep the molten puddle the same diameter at all times. This keeps the weld or *bead* the same width throughout its length. Start this bead 1/2 inch from the near edge of the plate being welded and proceed in a straight line parallel to this edge.

    Note: The width of the bead is directly related to the thickness of the plate being welded. The accepted standard for welds in aircraft tubing and light sheet metal requires the weld to be six times as wide as the thickness of the metal.

5. After the weld has been completed, examine it for uniformity of width and smoothness of appearance. Turn the plate over and examine the bottom for uniformity of *penetration.*

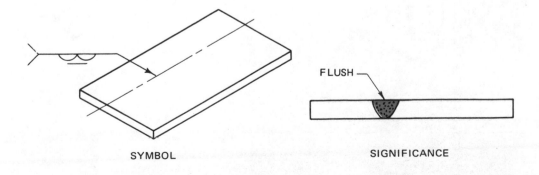

SYMBOL   SIGNIFICANCE

Fig. 14-1 Bead weld

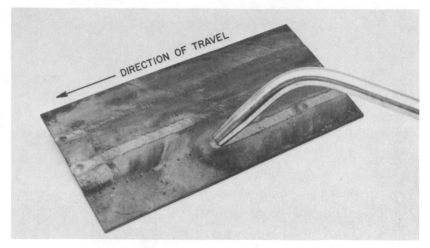

Fig. 14-2 Running a bead

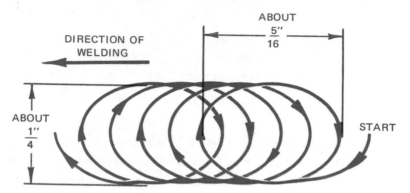

Fig. 14-3 Torch manipulation. Note: right hand operator

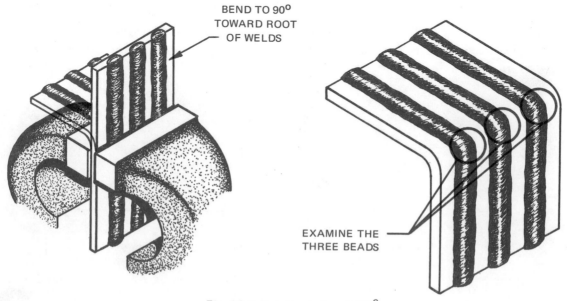

Fig. 14-4 Bending test weld 90°

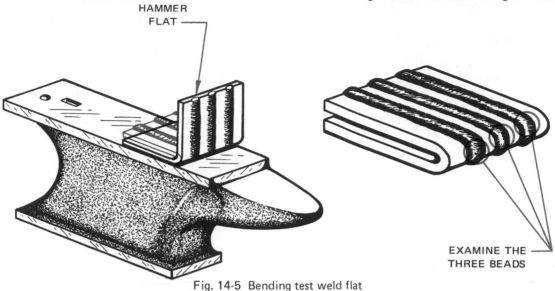

HAMMER
FLAT

EXAMINE THE
THREE BEADS

Fig. 14-5 Bending test weld flat

Note: The term penetration refers to the depth to which the parent metal is melted and fused. Fusion is the important factor. It is possible to obtain complete melting of the base metal and less than complete fusion. The welder cannot expect to produce a welded joint equal in strength to the base metal if the fusion is less than 100 percent.

6. Make more beads on the same plate parallel to the first bead and 1/2 inch apart. Vary the angle the flame makes with the work for each new bead. Observe the finished welds for appearance and penetration.

7. Continue to make more beads on additional plates until good appearance and penetration are attained. Manipulate the flame to obtain better results. The simplest manipulation is to rotate the flame in a small circle with a clockwise motion so that the flame is alternately closer and farther away from the work, figure 14-3. The frequency and length of these cycles give the welder added control of the amount of heat applied to the work. In most cases, a better appearing weld results when the flame is manipulated in this way.

8. Run a bead on another plate, using a neutral flame. Then adjust the flame to carburizing and run a bead parallel to the first and about 1/2 inch from it. Note that the sparks coming from the molten puddle tend to break into bushy stars. Notice the cloudy appearance of the molten puddle.

   Note: When steel is heated in a carbon rich atmosphere, such as that produced by a carburizing flame, it tends to absorb the carbon in direct proportion to the temperature and the amount of carbon present. This carbon combines with the steel to form the hard, brittle substance known as iron carbide.

9. Adjust the flame to highly oxidizing and run a third bead parallel to the second and 1/2 inch from it. Note that the molten puddle is violently agitated and that the molten iron oxide has an incandescent frothy appearance. Iron oxide is a hard, brittle, low-strength material of no structural value. Note, also, that the oxidized bead is much narrower than either of the others.

10. Cool the finished test plate and grasp it in a vise across the center of the three welds. Bend this plate 90 degrees toward the root of the welds, figure 10-4. Notice that the carburized bead has cracked and most of the oxidized material has fallen from the oxidized bead.

11. Hammer the test plate on an anvil until it is bent flat upon itself, figure 10-5. Notice that while the carburized and oxidized beads have cracked, the neutral bead has bent as much as the original material with no indication of failure.

REVIEW QUESTIONS

1. How does uniformity of procedure affect appearance of the finished weld?

2. What effect does flame angle have on penetration?

3. What effect does bead width have on fusion?

4. Does the force of the flame blow the molten puddle along the plate or does the molten puddle have to follow the direction the flame takes?

5. What type of flame is best for oxyacetylene welding operations? Why?

6. Why is the bead produced by the oxidizing flame narrower than that produced by the other two flames?

UNIT 15   MAKING BEADS WITH WELDING ROD

In most oxyacetylene welding, additional metal is added to the weld by melting a filler rod into the puddle to produce a stronger weld. These rods are available in various diameters and materials.

The use of the filler rod requires the operator to manipulate not only the torch but also the rod. The proper coordination of the torch and the rod is necessary for the production of good welds. This unit provides an opportunity for practicing this manipulation and observing the results.

MATERIALS

16- or-18-gage steel plate, approximately 4 in. X 9 in.
3/32-inch diameter steel welding rod
Airco® #1 or #2 welding tip or equivalent

PROCEDURE

1. Light the torch and adjust the flame to neutral.

2. Melt the base metal near one end of a plate until a puddle of the proper size is obtained as in unit 14.

3. Place the welding rod in the puddle, making sure the rod is aimed in the direction of travel of the weld, figure 15-2.

4. Proceed with the weld, making sure the welding rod and the tip of the torch make the correct angles with the work. Attempt to make a straight, uniform bead parallel to the edges of the base metal.

5. Make more welds in this manner but vary the angle that the rod makes with the work. Note the effect on the height of the bead.

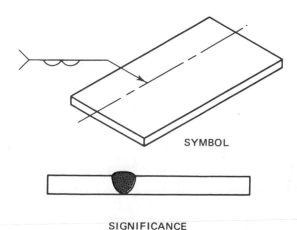

SYMBOL

SIGNIFICANCE

Fig. 15-1 Bead weld with welding rod

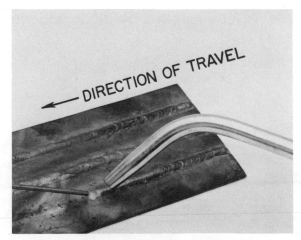

DIRECTION OF TRAVEL

Fig. 15-2 Running a bead with filler rod

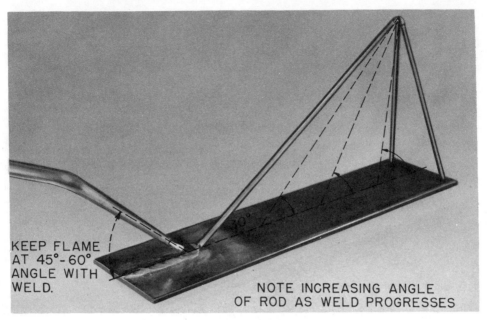

KEEP FLAME AT 45°-60° ANGLE WITH WELD.

NOTE INCREASING ANGLE OF ROD AS WELD PROGRESSES

Fig. 15-3 Welding with fixed rod

6. Obtain another plate and set up the plate and rod according to figure 15-3.

   Note: The rod should be twice as long as the legs and welded to them. This produces a starting angle of 30 degrees between the rod and the welding line. As the weld progresses and the rod melts, this angle gradually becomes greater. Observation of the finished weld shows the effect the rod and angle have on the height of the finished bead.

7. Make the weld and observe that in this case, the rod manipulation is not necessary to make a weld with good appearance.

8. Obtain more plates and rods and make more welds; but, as the welding progresses, dip the rod end into the molten puddle with a regular rhythm, figure 15-4. Try one second in the puddle and one second out, and then increase the rhythm until the dipping action is rather rapid.

9. Observe the effect this dipping action has on bead height and uniformity.

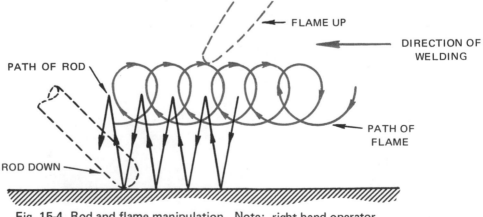

FLAME UP

DIRECTION OF WELDING

PATH OF ROD

PATH OF FLAME

ROD DOWN

Fig. 15-4 Rod and flame manipulation. Note: right hand operator

REVIEW QUESTIONS

1. What effect does rod angle have on the finished weld?

2. What effect does rod manipulation have on the weld?

3. What effect does uniformity of flame manipulation have on the weld?

4. What is wrong if the weld surface is flat in appearance?

5. What causes washout at the start and finish of a weld?

UNIT 16 TACKING LIGHT STEEL PLATE AND MAKING BUTT WELDS

When making a butt weld, the metal expands as heat is applied and contracts as it cools. This may distort the metal and cause an unsatisfactory job.

To avoid such distortions, several precautions may be taken. One of the most common is to *tack* the two pieces in position. Tack welds are small temporary welds to hold the work in place and control the distortion. A skilled operator must know how to place tack welds, and what effect they have on the finished job.

MATERIALS

Two pieces of 16- or 18-gage mild steel plate, 1 1/2 to 2 in. X 9 in. each
3/32-inch diameter steel welding rod
Airco® #1 or #2 welding tip or equivalent

PROCEDURE

1. Place the plates in position on the welding bench as indicated in figure 16-2.

2. Make tack welds approximately every two inches from right to left. The distance between tacks may be greater for thicker plates.

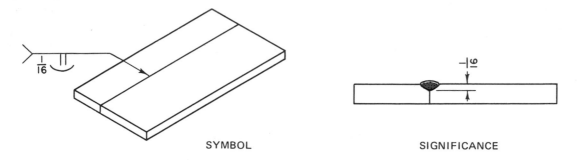

SYMBOL   SIGNIFICANCE

Fig. 16-1 Butt weld

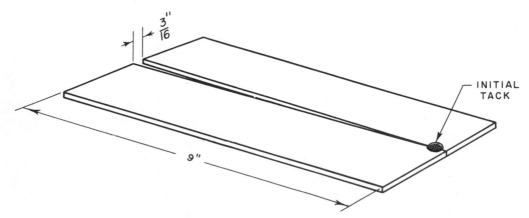

Fig. 16-2 Positioning plates for tacking

46

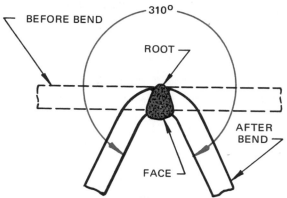

Fig. 16-3 Bend test for butt weld

3. Make the butt weld in much the same manner as in unit 15. However, the weld must be straight and the center of the bead must be on the exact center of the joint.

4. Examine the finished weld for uniformity; inspect the reverse side for penetration.

   Note: On light metal, penetration should be complete from one side of the plate. If this penetration is not obtained, secure more plates and make additional butt welds with wider beads. Practice this until penetration is complete. This happens when the width of the bead is about six times the thickness of the plate.

5. Test the butt welds, as shown in figure 16-3, by holding the finished weld in a vise with the centerline of the weld 1/8 inch above the jaws. Hammer the plate toward the face of the weld. A good weld shows no evidence of root cracks.

6. Obtain two more plates and tack both ends. Then try a third tack weld midway between the first two. Observe the effect of this procedure on plate alignment and ease of tacking.

REVIEW QUESTIONS

1. What happens if the plates are placed in contact for their entire length and then tacked?

2. What effect does plate thickness have on plate spacing?

3. What effect does weld width have on penetration?

4. What effect does tacking only the ends of the joint have on plate alignment during welding?

5. What is penetration?

UNIT 17   OUTSIDE CORNER WELDS

The welded shape in this unit is easily tested by a simple method to determine the quality of the fusion. Welds may be tested to discover poor welds and errors in the procedure used.

MATERIALS

Two pieces of 16- or 18-gage steel plate, 1 1/2 to 2 in. X 6 in. each
Airco® #1 or #2 welding tips or equivalent

PROCEDURE

1. Set up and tack the plates every two inches starting at the end.

2. Weld the plates by placing the flame on the work so that it is split by the sharp corner of the assembly, figure 17-4.

3. Make the weld, trying at all times to make a smooth uniform bead, figure 17-4.

4. Examine the finished bead for uniformity and complete penetration.

5. Check the finished bead by placing the assembly on an anvil and hammering the bead until the plates lie perfectly flat, figure 17-5. Examine the underside for cracks and lack of fusion.

6. Weld more joints of this type, varying the size of the puddle until complete penetration is obtained.

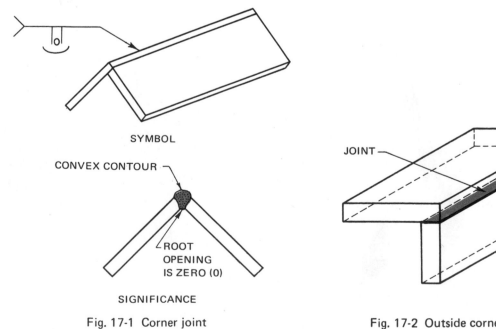

SYMBOL

CONVEX CONTOUR

ROOT OPENING IS ZERO (0)

SIGNIFICANCE

Fig. 17-1 Corner joint

JOINT

Fig. 17-2 Outside corner weld

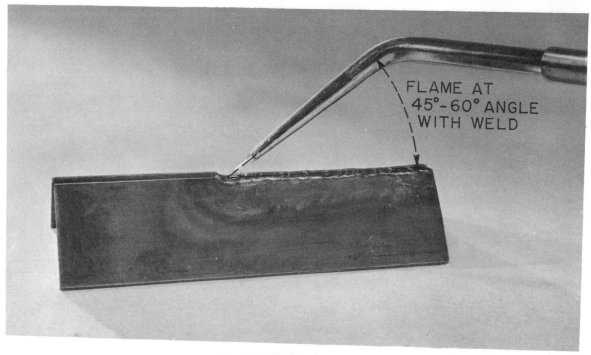

FLAME AT
45°-60° ANGLE
WITH WELD

Fig. 17-3 Welder's eye view

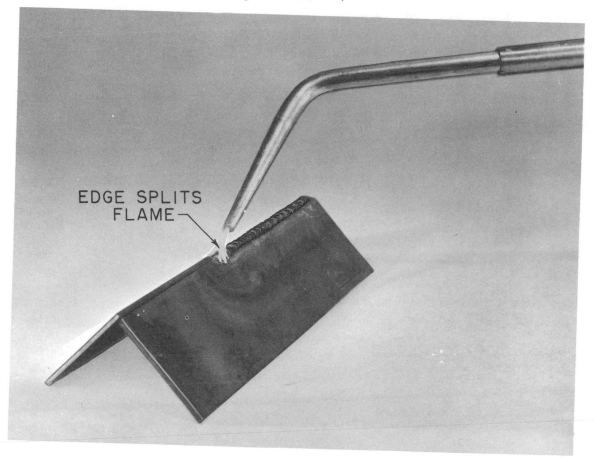

EDGE SPLITS
FLAME

Fig. 17-4 Corner Weld

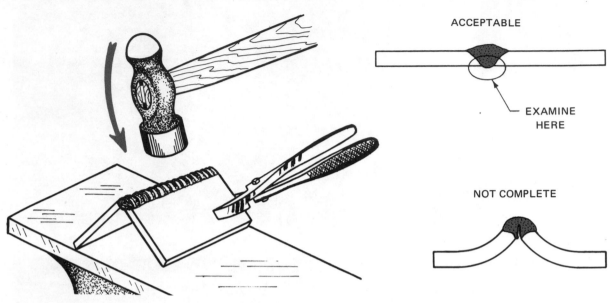

ACCEPTABLE

EXAMINE
HERE

NOT COMPLETE

Fig. 17-5 Testing the weld

Note: When welding mild steel with a neutral flame, one way to check for complete penetration is the number of sparks coming from the molten puddle. Very few sparks are produced when the penetration is not complete, and large numbers of sparks are noted when the penetration is excessive. Good welders learn to use this indication to insure proper welds.

REVIEW QUESTIONS

1. In hammer-testing this type of joint, what color is the break if the penetration is not complete?

2. How does the strength of a properly made outside corner weld compare with the base metal?

3. What effect does too much penetration have on the appearance of the finished bead?

4. Is the angle of the torch tip critical, when making this weld? Why?

5. What is the function of a tack weld?

UNIT 18   LAP WELDS IN LIGHT STEEL

The lap weld introduces some new difficulties from possible uneven melting of the two lapped plates. This type of welded joint illustrates the importance of the proper distribution of heat on the surface to be welded. One part of the joint may be melting while the other part may be far below melting temperature. This uneven heating prevents good fusion.

A simple test of the welded lap joint shows the quality of the weld. This shows the student where more practice is needed, and which welding procedures need to be changed to produce a better weld.

MATERIALS

Two pieces of 16-gage mild steel plate,
   2 in. X 6 or 9 in. each
3/32-inch diameter steel welding rod
Airco® #3 welding tip or equivalent

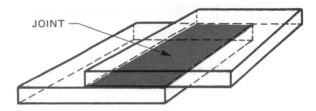

Fig. 18-1 Lap joint

PROCEDURE

1. Place the plates on the welding bench in the position shown in figure 18-3.

    Note: Tack weld the two pieces at the ends so the tacks will not interfere with the weld.

2. Weld the plates, holding the rod and flame as shown in figure 18-3.

    Note: The edge of a plate melts more readily than the center of the plate. Therefore, in this weld there is a tendency for the top plate to melt back too far. This is overcome by placing the rod in the puddle as shown in figure 18-4. The rod is tilted slightly toward the top plate. The rod then absorbs some of the heat and eliminates excessive melting of the top plate.

3. Examine the finished weld for uniformity and for excessive melting of the top plate.

4. Place the weld in a vise, figure 18-5, and test by hammering the lapping plate until it forms a T with the bottom plate.

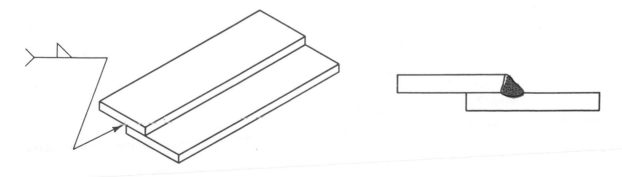

Fig. 18-2 Lap weld

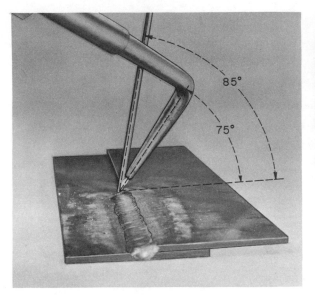

Fig. 18-3 Making the lap weld

Fig. 18-4 Welder's eye view

Fig. 18-5 Testing the weld

5. Examine the root of the weld for complete penetration.

6. Practice this type of joint until good surface appearance is obtained and root penetration is complete.

REVIEW QUESTIONS

1. Where should the greatest amount of heat be directed in the lap weld?

2. What effect does the position of the rod in the puddle have on the melting of the lapping plate?

3. What is the relative position of the top and bottom of the molten puddle while the weld is being made?

4. Are the rod and torch angles more or less critical in this job than in the previous jobs?

UNIT 19 TEE OR FILLET WELDS ON LIGHT STEEL PLATE

A tee or fillet weld in light steel plate provides experience in welding two steel plates set at right angles to each other. The angle of the flame to each of the two plates is important. The positions of the puddle and the rod also have an important effect on the quality of the weld. This job presents a new problem for the beginner — the possibility of *undercutting*.

MATERIALS

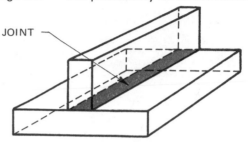

JOINT

Fig. 19-1 Tee joint

Two pieces of 16- or 18-gage steel plate,
   1 1/2 or 2 in. X 9 in. each
3/32-inch diameter steel welding rod
Airco® #2 or #3 welding tip or equivalent

PROCEDURE

1. Set up and tack the plates as shown in figure 19-4.

2. Establish the size and shape of the weld.

3. Proceed with the welding, and pay particular attention to the following points:

   a. The centerline of the flame should make an angle of 45 degrees or less across the bottom plate.

   b. The angle the flame should make with the weld centerline varies from 60 degrees to 80 degrees. For thicker plate the flame should be pointed more directly into the weld.

   c. The puddle of molten metal should be positioned so that the bottom of the puddle is slightly ahead of the top. This is done by rotating the flame in a clockwise direction so the flame follows an oval path.

   d. The rod is usually placed near the top of the puddle so that it comes between the flame and the upstanding plate. In this position, the rod absorbs some of the heat and prevents excessive melting (burning through), or undercutting the vertical leg.

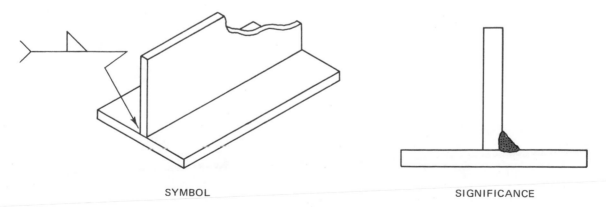

SYMBOL

SIGNIFICANCE

Fig. 19-2 Tee weld

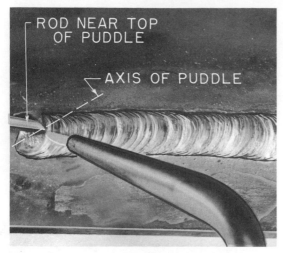

Fig. 19-3 Welder's eye view

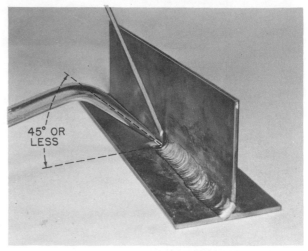

Fig. 19-4 Making the fillet weld

Note: Undercutting may be encountered, figure 19-5. This is an absence of metal along the top edge of the weld. It is caused by too much heat or poor rod movement. It should be avoided at all times.

4. Check the finished fillet weld by placing the assembly on an anvil and hammering the upstanding leg flat toward the face of the weld, figure 19-6. Examine for cracks in the root of the weld.

5. Make more joints of this type until smooth uniform welds are made. It should be possible to bend these welds 90 degrees in either direction without cracking.

6. After a fillet weld has been made on one side of the assembly, make a weld on the opposite side.

   Note: The first weld has produced oxide or scale on the reverse side. This is removed easily by playing the flame rapidly back and forth along the back surface of the joint. The flame causes the oxide to expand and pop from the surface.

CAUTION: When using the flame descaling procedure, extreme care must be used to protect the eyes and skin from burns caused by the hot, flying scale.

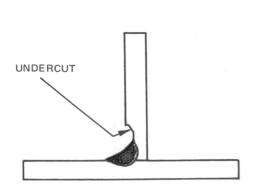

Fig. 19-5 Undercutting

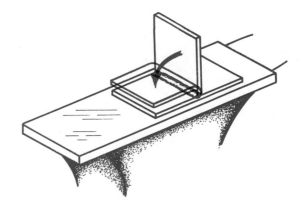

Fig. 19-6 Testing the fillet weld

REVIEW QUESTIONS

1. Is it possible to develop the full strength of the joint when undercutting is present? Explain.

2. What effect does flame and rod movement have on root penetration and appearance of the weld?

3. When making fillet welds on both sides of the joint, how does the amount of heat required for the second weld compare to the first? Why?

4. What effect does melting the excess oxide on the plates and fusing it with the second weld (in step 6) have on the finished bead?

5. Define undercutting.

UNIT 20   BEADS OR WELDS ON HEAVY STEEL PLATE

Although the principles of welding on heavy steel plate are the same as with lighter plate, the problems are greater because more heat is required. To distribute this heat properly, attention must be paid to torch motion and flame angle.

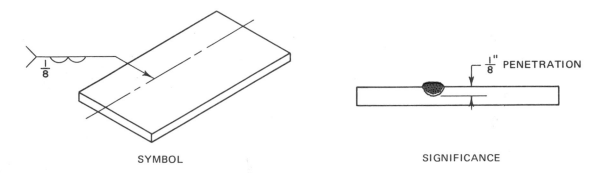

SYMBOL   SIGNIFICANCE

Fig. 20-1 Bead on heavy steel plate

MATERIALS

3/16-inch thick mild steel plate, 4 in. X 9 in.
1/8-inch diameter steel welding rod
Airco® #5 or #6 welding tip or equivalent

Fig. 20-2 Torch motion

PROCEDURE

1. Adjust the flame to neutral.

2. Apply the flame to the work with the tip at an angle of 75 to 80 degrees along the line of the weld.

3. Weld a bead the length of the plate as in unit 15. The flame should be moved in a half-moon weave to produce a weld of adequate width and depth of penetration, figure 20-2. Make the bead 1/2 inch to 5/8 inch wide.

4. Observe the finished bead for appearance, particularly the spacing of ripples, edge of fusion, and penetration.

5. Run more beads on the same plate by varying the angle that the tip makes with the work with each bead.

6. Observe these beads for uniformity. Determine the effect of too little or too great a tip angle on the penetration of the base metal and the appearance of the finished weld.

    Note: This job requires a much larger tip size than any used in previous jobs. As a result, the gas consumption rises very rapidly with a corresponding rise in the hourly cost of operation. The gas should be shut off as soon as the welding is completed so that the cost of operation can be kept down.

REVIEW QUESTIONS

1. What effect does flame angle have on the size and shape of the puddle and the bead ripples?

2. Draw a sketch of the puddle when the flame angle is too small.

3. What effect does flame angle have on penetration in heavy plate?

4. Is the tip manipulation the same for heavy plate as it is for light plate?

5. Does penetration become more of a problem on heavy plate? How is it helped?

UNIT 21 MANIPULATION OF WELDING ROD ON HEAVY STEEL PLATE

Considerable practice is required to develop skill in manipulating the welding rod and flame in the molten puddle when welding heavy steel plate. The relationships of the rod, flame, and puddle are particularly important. This unit provides practice in manipulating all three.

MATERIALS

3/16- or 1/4-inch thick mild steel
   plate, 4 in. X 9 in.
1/8-inch diameter steel welding rod
Airco® #5 or #6 welding tip or equivalent

Fig. 21-1 Torch and rod motion

PROCEDURE

1. Apply a neutral flame to the work as in unit 20.

2. The molten puddle should be 1/2 inch to 5/8 inch wide.

3. Weld as in unit 20 except that both the rod and the flame should be moved alternately. In other words, the rod and flame should be moved so that they are on opposite sides of the molten puddle at all times, figure 21-1.

4. Inspect the finished weld for appearance.

5. Make more parallel welds on the same plate, but vary the angle of the rod and the flame. Observe the effect that this variation has on the height of the weld, the depth of penetration, and the face of the weld.

6. Compare these beads or welds with those made in unit 20.

REVIEW QUESTIONS

1. What is the advantage of moving the flame and rod, rather than the flame alone?

2. Which weld has the more uniform appearance?

3. Can a weld bead on heavy plate be too large?

4. Will the surface appearance of the bead be more uneven on heavy plate than light plate welding?

5. On heavy plate is the penetration the same as, less than or more than that on light plate if the correct tip is used?

UNIT 22   BUTT WELDS ON HEAVY STEEL PLATE

Butt welding heavy steel plate is a basic welding operation. The quality of the weld is determined by: the positioning of the plates, tacking the plates, preparation of the edges, and the fit of the plates to each other. All of these factors also affect the ease with which the weld may be made.

Besides providing another opportunity to acquire more skill in butt welding, this unit points out the importance of plate edge spacing.

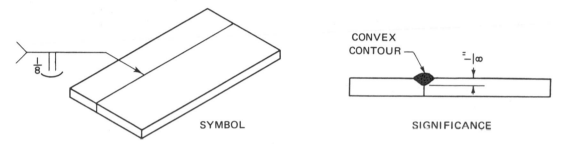

Fig. 22-1 Butt weld on heavy steel plate

MATERIALS

Two pieces of 3/16-inch or 1/4-inch thick mild steel plate, 1 1/2 to 2 in. X 9 in. each
1/8-inch diameter steel welding rod
Airco® #5 or #6 welding tip or equivalent

PROCEDURE

1. Align the plates and tack as in making butt welds in thin plate, unit 16. The distance between tacks may be greater here than on the thinner material.

2. Proceed with welding as in unit 21.

3. Obtain more plates and tack them so that they are spaced 1/16 inch apart for one pair, 3/32 inch apart for another pair, and 1/8 inch apart for a third pair, figure 22-2.

   Note: When the plate edges are touching as in step 1, the joint is called a *closed square butt joint.* When they are spaced as in step 3, the joint is an *open square butt joint.*

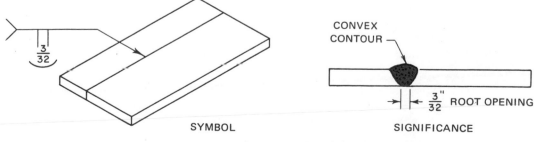

Fig. 22-2 Open square butt joint

4. Weld each pair of plates.

5. Examine and compare the finished butt joints.

REVIEW QUESTIONS

1. What is the relationship between plate thickness and distance between tacks?

2. What effect does plate edge spacing have on:

   a. Penetration?

   b. General appearance of the welds?

3. Draw a sketch of a cross section of a closed square butt joint.

4. Draw a sketch of a cross section of a open square butt joint.

5. Why is there a tendency for the open square butt joint to melt away at the plate edges?

UNIT 23   LAP WELDS ON HEAVY STEEL PLATE

The procedure for making a lap or fillet weld on heavy steel plate is much the same as that required for making this weld on light plate.

In the job performed in this unit, more metal must be deposited because of the thicker plate. Since the heat must cover a larger area, more torch movement is involved. In addition, the larger molten puddle requires more rod movement.

MATERIALS

Two pieces of 3/16 or 1/4-inch thick steel plate, 2 in. X 9 in. each
1/8-inch diameter steel welding rod
Airco® #5 or #6 welding tip or equivalent

PROCEDURE

1. Align the plates so that they lap approximately halfway in the long direction, figure 23-1.

   Note: Tack weld the two pieces at the ends as in unit 18.

2. Weld the plates, but point the flame toward the joint more than when lap-welding light steel plate, figures 23-2 and 23-3. Keep the flame pointed more toward the top plate. The tendency to overmelt the top plate is much less than in welding light plate.

3. As the welding proceeds, rotate the flame clockwise so that the bottom of the molten puddle is slightly ahead of the top. Try varying the position of the rod in the molten puddle.

4. Obtain more plates and repeat this type of joint; but alternate the flame and rod in the puddle in much the same manner as in the butt joint. Manipulate the flame and rod to keep the bottom of the puddle slightly in advance of the top.

5. Make more joints of this type. In each weld, vary the amount that the bottom of the puddle leads the top.

6. Inspect the finished welds for appearance.

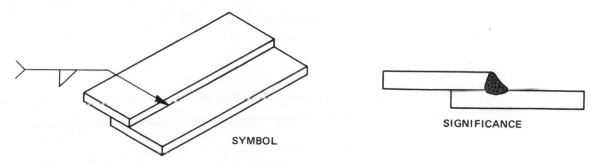

SYMBOL

SIGNIFICANCE

Fig. 23-1 Lap weld

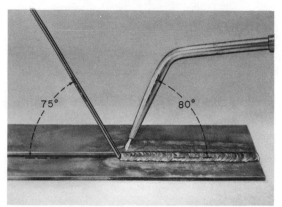

Fig. 23-2 Welder's eye view

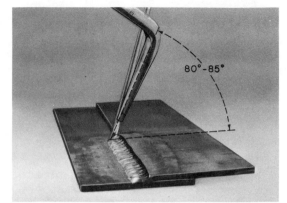

Fig. 23-3 Lap weld on heavy steel plate

7. Break these joints in a vise and examine the welds. Inspect for complete fusion of both plates. Note the size of the crystals or grain of the weld metal in the break.

8. In a piece of steel about the same size, make a saw cut part way through, so it breaks when bent. Compare the grain structure or crystal size of this break with the grain structure of the broken weld metal.

REVIEW QUESTIONS

1. What effect does the tip angle have on the appearance of the finished bead?

2. What effect does the tip angle have on the fusion of the two plates?

3. What effect does the heat of fusion have on the size of the grain structure in the weld and nearby metal?

4. Does undercut become more of a problem with heavy plate? Why?

5. Is it possible for a lap weld to look good and not be strong? Explain.

UNIT 24 FILLET OR TEE JOINTS IN HEAVY STEEL PLATE

Fillet or tee joints are welded in much the same way as the lighter plates in unit 19. However, changes must be made in the flame angle, rod position, and the molten puddle to produce a good weld.

MATERIALS

Two pieces of 3/16- or 1/4-inch thick mild steel plate, 2 in. X 9 in. each
1/8-inch diameter steel welding rod
Airco® #5 or #6 welding tip or equivalent

PROCEDURE

1. Position the plates so that the upstanding plate is about in the center of the flat plate. Tack the plates so that the upstanding leg makes an angle of exactly 90 degrees with the bottom plate.

2. Make the weld, using much the same technique as when making the weld in unit 23. The flame angle and rod position are very important and must be correct to avoid under-cutting the upstanding leg, figure 24-2.

3. Check the angle between the vertical and horizontal plates after the weld has cooled.

4. Make more joints, tacking the plates at slightly varying angles. This will allow for shrinkage as the weld cools. Check the angle between the vertical and horizontal plates after the weld has cooled.

5. Make more joints of this type, varying the flame angle, rod position, and amount of lead of the bottom of the puddle.

   Note: The size of the legs of the weld must equal the thickness of the plate being welded if the joint is to be full strength.

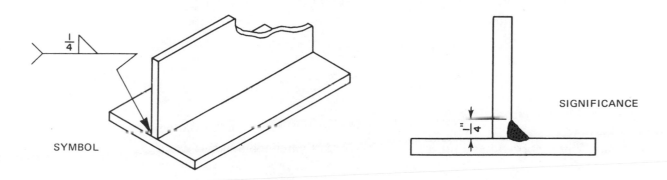

SYMBOL

SIGNIFICANCE

Fig. 24-1 Fillet weld

63

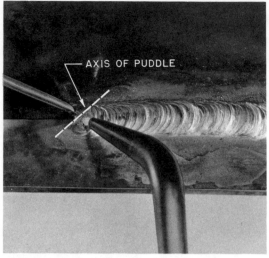

Fig. 24-2 Fillet weld on heavy steel plate

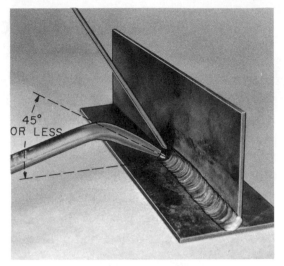

Fig. 24-3 Welder's eye view

6. Break the joints by placing the bottom plate in a vise and bending the upstanding leg toward the weld face.

7. Examine the welds for fusion and penetration, especially at the root of the weld.

REVIEW QUESTIONS

1. What effect does flame angle have on the appearance of the finished bead?

2. What effect does flame angle have on the tendency to undercut?

3. What effect does rod position in the molten puddle have on appearance and the tendency to undercut?

4. How are the plates set up so the final angle is exactly 90 degrees, after shrinkage? Make a cross section sketch of this joint as set up ready for welding.

5. Why is correct hand protection so important when making this weld?

UNIT 25   BEVELED BUTT WELD IN HEAVY STEEL PLATE

Complete fusion and penetration are of great importance to the welding operator. Joints which have been V'd or beveled make penetration in heavy plate much easier. This unit provides practice in flame cutting, flame and rod movement, and multilayer welding.

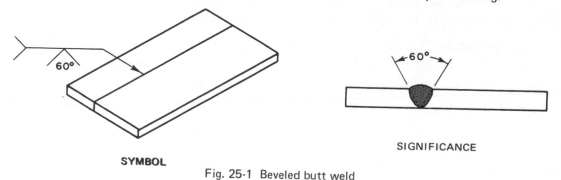

SYMBOL

SIGNIFICANCE

Fig. 25-1 Beveled butt weld

MATERIALS

Two pieces of 5/16- or 3/8-inch thick mild steel plate, 3 in. X 9 in. each
1/8-inch diameter steel welding rod
Airco® #7 welding tip or equivalent

PROCEDURE

1. Flame-cut one edge of each of the plates so that the included angle is 60 degrees.

2. Align and tack the plates with 3/32-inch root opening, figure 25-1.

3. Make the first weld in the bottom of the groove. Fusion between the plates should be complete. The flame should make an angle of 60 to 75 degrees with the work, figure 25-2.

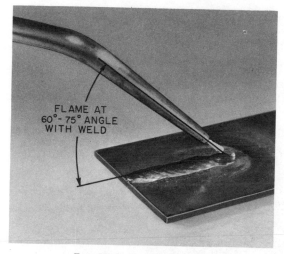

FLAME AT
60°- 75° ANGLE
WITH WELD

Fig. 25-2 Running a bead

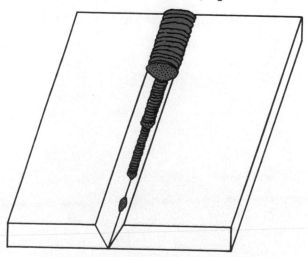

Fig. 25-3 Multilayer welding

If there is a tendency to burn through at the bottom of the V, try a smaller flame angle. If the bead is irregular and the fusion is poor, try a greater flame angle.

4. Apply the second bead as shown in figure 25-3. Increase the flame angle and weave the torch slightly. Apply only enough rod to make a weld that is flat or somewhat concave.

5. Make the third bead, weaving the flame and rod alternately, as in step 3, unit 21. Control of the flame is important at this point so the fusion is complete along the top edges of the V. However, the melting must not be so great as to make the weld too wide. The weld should not be over 1/8 inch wider than the top of the opening.

6. Make two saw cuts across the finished weld to produce a sample 1 1/2 inch wide. Examine these sawed surfaces for complete fusion and absence of gas pockets or holes in the weld.

7. Break this sample by hammering it toward the face of the weld. Examine the break for lack of fusion, oxide in the weld, and gas pockets.

REVIEW QUESTIONS

1. What effect does too small a flame angle have on the first bead?

2. How can you overcome the tendency of the first and second beads to pile up or become convex?

3. How do the flame and rod motion, and the flame angle in this unit compare with those in unit 21?

4. How can poor fusion at the root of the weld be corrected?

5. Is the root spacing of the joint critical for good penetration? Why?

UNIT 26   BACKHAND WELDING ON HEAVY STEEL PLATE

All of the welds made so far have been made by the *forehand* method with the flame pointing in the direction of welding. In the *backhand* method, the torch and rod are held in the same position but the flame points opposite the direction of welding. The welding flame is directed at the completed portion of the weld. The welding rod is placed between the completed weld and the flame.

MATERIALS

Two pieces of 1/4-inch thick mild steel plate, 3 in. X 9 in. each
1/8-inch diameter steel welding rod
Airco® #7 welding tip or equivalent

PROCEDURE

1. Align and tack the plates as shown in figure 26-1.

2. Start a puddle at the end of the joint and proceed with the weld by the backhand method. Use a weaving motion similar to that in units 21 and 25.

3. As the weld progresses, observe the bead and alter the flame angle until the weld has good appearance.

4. Make joints of this type until welds are produced with uniform ripples and complete fusion throughout the entire length.

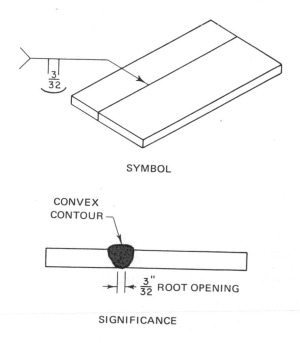

SYMBOL

CONVEX CONTOUR

$\frac{3}{32}$" ROOT OPENING

SIGNIFICANCE

Fig. 26-1 Butt joint in heavy plate

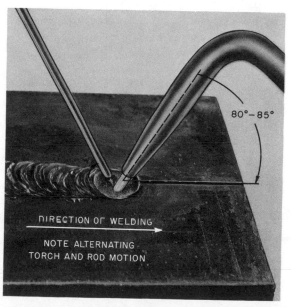

80°–85°

DIRECTION OF WELDING

NOTE ALTERNATING TORCH AND ROD MOTION

Fig. 26-2 Welder's eye view: backhand welding

5. As these welds proceed, observe the amount of penetration and the width of the beads. Compare these welds with those made in unit 21.

6. Make two saw cuts across the finished weld to produce a sample 1 1/2 inches wide. Examine these sawed surfaces for complete fusion and absence of gas pockets in the weld.

7. Break this sample by hammering it toward the face of the weld. Examine the break for lack of fusion, oxides in the weld, and gas pockets.

REVIEW QUESTIONS

1. How does the width of a backhand weld compare with the width of a forehand weld on a plate of similar thickness?

2. How does the flame angle compare with forehand welding?

3. Sketch a lengthwise cross section of this type of weld showing the finished bead, the rod, and flame penetration.

4. What is the advantage of backhand welding over forehand?

5. Which type of welding is faster, forehand or backhand?

UNIT 27   BACKHAND WELDING OF BEVELED BUTT JOINTS

Backhand welding of beveled butt joints is used to great advantage in oxyacetylene welding of steel pipe. This unit provides practice in making this type of joint. Although this joint is not easy to make, it is much less difficult than a similar one in pipe. This joint should be mastered before pipe joints are welded.

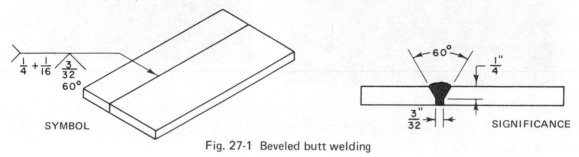

Fig. 27-1 Beveled butt welding

MATERIALS

Two pieces of 5/16- or 3/8-inch thick
   mild steel plate, 3 in. X 9 in. each
3/16-inch diameter steel welding rod
Airco® #7 or #8 welding tip or equivalent

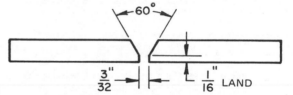

Fig. 27-2 Cross section of beveled plates

PROCEDURE

1. Bevel the plates as indicated in figure 27-1.

2. If equipment is not available to produce the *land* shown in figure 27-2, bevel the plates to a feather edge and grind them to obtain the amount of land indicated. The land is the vertical part of the opening.

3. Align and tack the plates as shown.

4. Weld the root by the backhand method so that fusion is complete. This is done by using the proper flame angle and rod application. The flame angle must be steep enough to insure good penetration. The rod may have to be dipped in and out of the puddle to allow the heat to melt the root of the base metal.

5. Make the second pass, using the motion described in unit 21. Be sure the bead is completely fused, and that it is not over 1/8 inch wider than the top of the original groove, figures 27-3 and 27-4.

6. Cut a section from this weld and examine it for penetration and fusion.

7. Weld additional plates with one pass only on each bead.

8. Cut a section from each of these joints and inspect them as in step 6. Check especially for penetration, fusion, and bead appearance.

Fig. 27-3 Welder's eye view

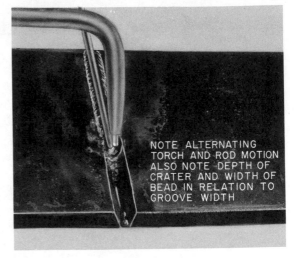

NOTE ALTERNATING TORCH AND ROD MOTION ALSO NOTE DEPTH OF CRATER AND WIDTH OF BEAD IN RELATION TO GROOVE WIDTH

Fig. 27-4 Beveled butt welding — backhand method

REVIEW QUESTIONS

1. What should be the shape of the surface of the root pass when making the joint at step 4?

2. Why does step 5 caution that the face of the bead must be kept narrow?

3. What is the difficulty in making this joint with a single pass?

4. Does it take longer to make the joint by the forehand method or the backhand method?

5. Will the welder feel more heat or less heat in backhand welding?

UNIT 28 BRAZING WITH BRONZE ROD

Brazing is a process in which metals are joined at a temperature greater than 800 degrees F. The base metal has a melting point at least 50 degrees F. higher than the filler rod. This indicates that:

- The base metal is not melted during this process.

- The joint is held together by the adhesion of the brazing alloy to the base metal rather than by cohesion. Cohesion takes place when the base metal and filler rod are fused.

- A brazed joint is bonded rather than welded.

This unit defines the techniques involved in oxyacetylene torch brazing using bronze filler rods. The action of flux and bronze during the brazing process will be examined.

MATERIALS

1/16- to 1/8-inch thick clean steel plates, 2 in. X 9 in. each
1/8-inch diameter bronze rod
Welding tip one size larger than for welding a similar plate
Suitable dry-type brazing flux

PROCEDURE

1. Place a piece of *clean* steel plate on the welding bench so that one end overhangs the bench.

    Note: The word clean refers to steel which has all the mill scale or iron oxide removed by either chemical or mechanical means. Mill scale and rust makes the production of strong joints either very difficult or impossible.

2. Sprinkle some flux on this plate. Apply the flame to the bottom side of the plate until the flux melts and flows over the plate.

3. Observe the color of the plate at this temperature. Also note that the flux flows freely. These two factors are the best guides to the proper brazing temperature.

4. Cool the plate and observe the metal under the flux. Compare the color of this metal to that of the unheated part of the plate. Note that the fluxed part of the plate is much whiter in color, indicating that the flux has cleaned the metal. The primary purpose of the flux is to chemically clean the surface for the brazing alloy. A secondary purpose is to protect the finished bead from the atmosphere during the brazing and cooling period.

5. Put some flux and a drop of brazing alloy on the end of an overhanging plate and heat as before. Observe that the flux melts first, then the alloy.

6. Move the flame about on the bottom of the plate. Note that the alloy flows freely in all directions as long as the flux is flowing ahead and cleaning the metal.

Note: The process of adhering a thin coating of bronze or some other metal to the surface of the base metal is called *tinning.*

7. Continue to apply heat. Note that the alloy starts to burn with a greenish flame, first in small spots and then in wider areas. At the same time, the alloy gives off a white smoke and leaves a white residue on the plate.

Note: This residue is the zinc being overheated to the point where it evaporates and burns, to form zinc oxide. This heating is harmful to the brazed joint because the alloy is changed by the removal of the zinc which is replaced to some extent by the zinc oxide.

CAUTION: Breathing of zinc oxide may cause the operator to become violently ill.

8. Place a drop of the brazing alloy on the end of another overhanging plate and heat as before but do not use flux. Note that the alloy does not spread over the plate as before. Instead, it tends to vaporize and burn when the melting point is reached. This indicates that a brazed joint cannot be made without a dry-type brazing flux or one of the paste-type fluxes.

REVIEW QUESTIONS

1. What is the major difference between a brazed joint and a welded joint?

2. It is possible to braze brass or bronze if the proper alloy is used. What two conditions determine whether the joint is brazed or welded?

3. How does the flux act as a guide to the temperature of the joint?

4. From observations of the flowing characteristics of this alloy, what precaution should be taken to insure that the bead does not become too wide?

5. Can brazing create a toxic atmosphere for the welder? Explain.

UNIT 29　RUNNING BEADS WITH BRONZE ROD

This job provides practice in flame movement so that the operator develops skill in making good beads of a given size and shape with bronze rod.

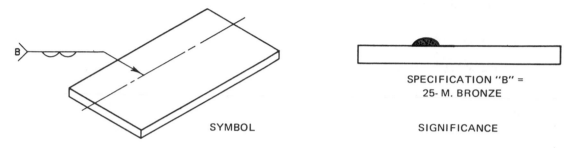

SYMBOL

SPECIFICATION "B" =
25- M. BRONZE

SIGNIFICANCE

Fig. 29-1　Bead with bronze rod

MATERIALS

16- or 20-gage steel plate, 2 in. X 9 in.
1/8-inch diameter bronze rod
Suitable brazing flux
Welding tip one size larger than for welding on similar plate

PROCEDURE

1. Adjust the flame according to instructions. The flame for brazing varies with the alloy being used from slightly carburizing through neutral to slightly oxidizing.

2. Heat the rod with the flame until the flux clings to the rod when it is dipped in the flux container.

3. Apply the flame to both the rod and the work until a drop of alloy is left on the work. Remove the rod from the flame and continue to heat the plate until the drop melts and flows over an area of about 1/2-inch diameter.

4. Start a bead lengthwise on the plate. Keep the rod close to the flame and move both the rod and the flame in a spiral. Both the rod and flame are alternately close to the work and far away.

    Note: Check the flame angle. It is usually less than that used to weld a plate of similar size.

5. Bring the rod and flame in contact with the work when they are on the downswing of the spiral motion.

6. Drag the rod in the direction of the brazing before removing it for the upswing. This draws the flux ahead of the molten alloy and speeds the cleaning process.

7. Continue with the brazing and note that the flux flows ahead of the alloy. When this no longer happens, dip the still hot rod in the flux and continue with the bead.

8. Inspect the finished bead for width, height, ripples, and for the white residue that indicates overheating.

9. Make more beads, using these variations in procedure:

    a. Change the size of circle made by the flame and rod.
    b. Apply the flame to the work for longer and shorter intervals of time.
    c. Increase and decrease the amount of flux on the rod.

10. Inspect the finished beads.

REVIEW QUESTIONS

1. What effect does too much or too little heat have on the appearance of the finished bead?

2. What effect does too much or too little flux have on the ease of brazing?

3. Is temperature control in the base metal more critical in brazing than in welding? Explain.

4. Is it possible to hold the inner cone of the flame farther away from the joint than when welding in order to make a narrower braze? Explain.

5. What is the color of the plate when it is at the proper temperature for brazing?

UNIT 30   SQUARE BUTT BRAZING ON LIGHT STEEL PLATE

This job gives practice in making butt brazes in light steel plate. Although fusion welding and brazing are two different processes, they have many points in common. Thus, it is possible to use much of the experience gained in welding to produce braze joints.

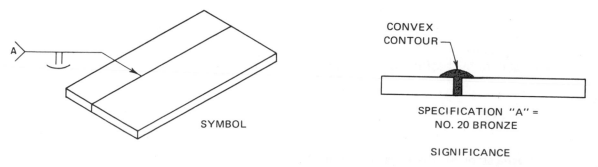

SYMBOL

CONVEX CONTOUR

SPECIFICATION "A" = NO. 20 BRONZE

SIGNIFICANCE

Fig. 30-1 Square butt braze

Two pieces 1/16- to 1/8-inch thick steel plate, 1 1/2 to 2 in. X 9 in. each
1/8-inch diameter bronze rod
Suitable dry brazing flux
Welding tip one size larger than for welding comparable steel plate

PROCEDURE

1. Align the plates and tack with bronze. Be sure the edges of the plates are in contact along the entire length of the joint.
2. Adjust the flame for the alloy being used.
3. Braze as in unit 25. Try to keep the bead narrow.
4. Hold the plates in a vise and bend them until they break, following the procedure for welded joints.
5. Examine the broken joint for evenness and depth of bond.
6. Tack more plates, spacing the edges slightly farther apart. Braze as before.
7. Break these plates and examine the results.

REVIEW QUESTIONS

1. What effect does plate edge spacing have on the depth of bond?

2. What is the procedure for getting a greater bond depth?

3. Is any tinning apparent on the reverse side of the joint brazed in this unit? If not, what conditions are necessary to obtain such tinning?

4. Is the square butt joint a good type of brazing joint? Why?

5. Is brazing stronger than fusion welding?

UNIT 31   BRAZED LAP JOINTS

This job develops manipulative skill in making brazed lap joints on steel plate. Some of the procedures, problems and difficulties encountered are the same as those in welding lap joints in light steel.

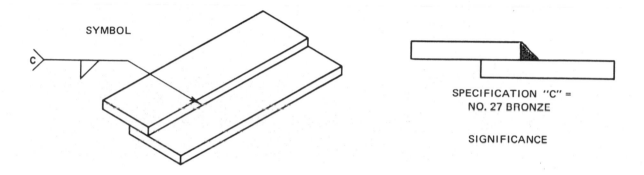

SYMBOL

C

SPECIFICATION "C" =
NO. 27 BRONZE

SIGNIFICANCE

Fig. 31-1 Brazed lap joint

MATERIALS

Two pieces of 1/16- to 1/8-inch thick clean steel plate, 2 in. X 9 in. each
1/8-inch diameter bronze rod
Suitable flux
Welding tip one size larger than for welding comparable plate

PROCEDURE

1. Lap the plates following the procedure outlined for welding a lap joint, unit 18.

2. Adjust the flame for the alloy being used.

3. Proceed with brazing and observe the tendency of the alloy to flow on the top plate.

4. Change the angle the flame makes with the line of brazing to correct the tendency to over braze the top plate, figure 31-2.

5. Continue to make these joints until enough skill is acquired to make brazed lap joints with a good appearance.

    Note: Try to make all brazed beads about the same width as a weld made on material of similar thickness. The tinning action of the alloy causes the braze to become too wide if flame movement is not carefully controlled.

6. Break and examine these joints using the same procedure as that used to break welded lap joints.

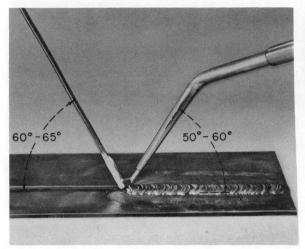

Fig. 31-2 Welder's eye view   Fig. 31-3 Brazed lap joint

REVIEW QUESTIONS

1. How critical is the flame angle as compared to that used when welding a similar joint?

2. What effect does holding the inner cone too far from the work have on the width of the finished bead?

3. When brazing a lap joint, is there a tendency for the alloy to flow between the plates?

4. How does the grain size of the break in the brazed joint compare with that of a welded lap joint?

5. What is the strongest type of brazed joint?

UNIT 32   BRAZED TEE OR FILLET JOINTS

This job helps the student obtain skill in the technique of making strong joints of good appearance using bronze filler metal.

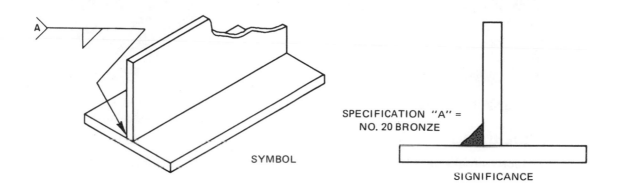

SYMBOL

SPECIFICATION "A" =
NO. 20 BRONZE

SIGNIFICANCE

Fig. 32-1 Brazed tee or fillet joint

MATERIALS

Two pieces of 1/16- to 1/8-inch thick clean steel plate, 2 in. X 9 in. each
1/8-inch diameter bronze rod
Flux
Welding tip one size larger than for welding similar plate

PROCEDURE

1. Set up the plates as for welding, unit 19, except that the tacks are made with bronze alloy.

2. Adjust the flame.

3. Proceed with brazing in much the same manner as when welding.

4. Observe the tendency of the bronze to tin the upstanding leg of the joint over large area.

5. Correct the flame angle and the distance of the inner cone from the work until this excessive tinning is overcome. Figure 32-3 shows the correct flame angle.

6. Bend the finished joint in the manner used to test the welded fillet joint. Examine the root of the joint for uniformity of bond.

7. Braze the opposite side of the joint. Note the difficulty in achieving a good bond on the upstanding leg.

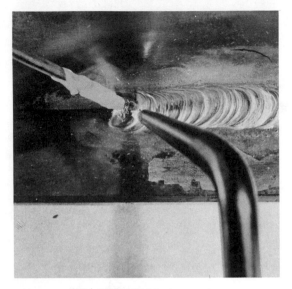

Fig. 32-2 Welder's eye view

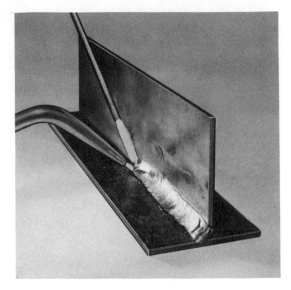

Fig. 32-3 Brazed fillet joint

REVIEW QUESTIONS

1. What factors make it difficult to obtain a good bond on the second side of the joint?

2. What preparation is necessary to obtain a good bond on the second side of the joint?

3. How does the flame for brazing the reverse side of the joint differ from that used on the first braze? Why is there a difference?

4. How should material be prepared for brazing?

5. What makes this joint more difficult to braze?

UNIT 33 BRAZING BEVELED BUTT JOINTS ON HEAVY STEEL PLATE

The lower temperatures used for brazing as contrasted with fusion welding make this process good for many jobs. Melting of the base metal is avoided, extensive preheating is not necessary, and expansion and contraction are not as severe problems as they are in fusion welding.

The student should acquire an understanding of brazing and will develop skill in this process. This unit provides an opportunity for brazing heavier steel than that used in previous units.

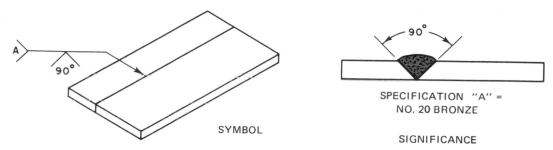

SYMBOL

SPECIFICATION "A" = NO. 20 BRONZE

SIGNIFICANCE

Fig. 33-1 Brazed beveled butt joint

MATERIALS

Two pieces of 1/4-inch thick steel plate, 2 in. X 6 in. each
1/8-inch diameter bronze rod
Flux
Welding tip one size larger than for welding similar plate

PROCEDURE

1. Prepare the plate edges so that they make an angle of 90 degrees when they are brought together, figure 33-1.

   Note: If the plates are flame-cut, lightly grind or file the cut edges until all oxides are removed.

2. Align and tack the plates as for welding, figure 33-2.

3. Braze the joint. Be sure the plate edges are tinned to the root of the joint. Be careful not to overheat the tip edges of the plate as the brazing proceeds.

4. Make more joints of this type but apply the alloy in two layers. Apply the second layer with a weaving motion, alternating the rod and flame in the same manner as in welding heavy plate. When making a multiple-pass braze, be sure that each bead, except the finished bead, is slightly concave and that the alloy tins well up on the sides of the joint, figure 33-3.

5. Vary the angle of the flame and observe the results.

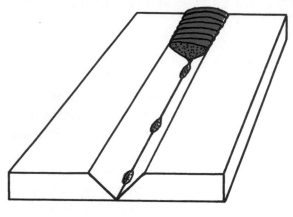

Fig. 33-2 Single-pass braze

Fig. 33-3 Multiple-pass braze

REVIEW QUESTIONS

1. What effect does the flame angle have on the penetration into the root of the groove?

2. What effect does the flame angle have on the appearance of the finished bead?

3. Why is the included angle for this joint 90 degrees instead of the 60 degrees used when welding a beveled butt joint?

4. Should the surface of the beads be flat, convex or concave?

5. Should the bottom sides of the material being brazed be cleaned? Why?

UNIT 34 BUILDING-UP ON CAST IRON

Worn areas of cast iron machine parts may be built up with bronze by the brazing process. The new surface may be machined if necessary. The resulting job is often as good as a new part.

Brazing on cast iron has advantages over fusion welding, largely because of the lower temperatures used. Brazing saves time, uses less gas, and involves less expansion and contraction of the metal. This unit gives practice in this process which is often called *bronze surfacing.*

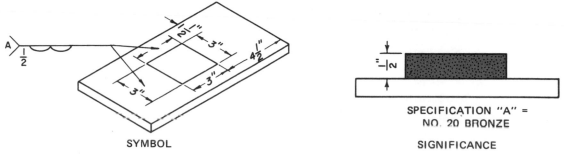

SYMBOL

SPECIFICATION "A" =
NO. 20 BRONZE

SIGNIFICANCE

Fig. 34-1 Built-up cast iron using bronze

MATERIALS

Cast iron
1/8-inch diameter bronze rod
Suitable flux
Welding tip one size larger than that indicated for welding metal of similar thickness

PROCEDURE

1. Clean the surface to be brazed. Remove all rust, scale, and oil.

2. Preheat the cast iron with the flame before attempting to braze.

3. Build up an area about 3 inches square with the bronze rod using beads 3/4 inch wide, figure 34-1. Make sure that each bead is fused into the preceding bead. Use a standard dry flux for this operation.

    Note: If a Hi-bond® cast iron brazing flux or its equivalent is available, use this in with the standard flux. Note the ease of tinning on the cast iron surface.

4. Make the first layer about 1/8 inch high. Apply a second bead over the first as soon as the work has been inspected and before the plate has a chance to cool.

5. Apply a third and fourth bead. Inspect each bead and correct any faults when making the next bead.

6. Grind one edge of a piece of cast iron 1/4 inch thick and 6 inches long until it is square and clean. Stand this plate on edge so that a bead can be applied to the ground edge, figure 34-2.

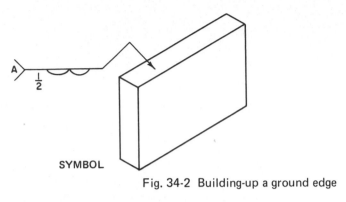

SYMBOL

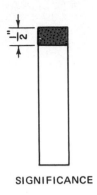

SIGNIFICANCE

Fig. 34-2 Building-up a ground edge

7. Apply a bead along this edge, moving the flame and rod so that the bronze flows to each side of the plate but does not overhang.

8. Inspect the finished bead for uniformity of thickness and ripples. Note the presence or absence of overhang.

9. Deposit additional beads over the first until the edge is built up to a height of 1/2 inch above the original plate, figure 34-2. Check for uniformity.

REVIEW QUESTIONS

1. How do the tinning characteristics of cast iron compare with those of the same thickness of steel?

2. Why is preheating of cast iron necessary?

3. Draw a sketch of the cross section of the bead and casting in step 8.

4. Can cast iron be too hot to braze? What happens?

5. Is there an advantage to using stringer beads in brazing instead of wide beads? Why?

UNIT 35   BRAZING BEVELED JOINTS ON CAST IRON

The quality of a brazed joint on cast iron is greatly affected by the surface preparation. The surfaces must be clean to allow good tinning and to get a strong bond. The two surfaces to be joined must have enough area to provide a strong joint.

Testing the brazed joint to destruction demonstrates some of the qualities of a good joint. This unit provides an opportunity to make and test brazed joints on cast iron.

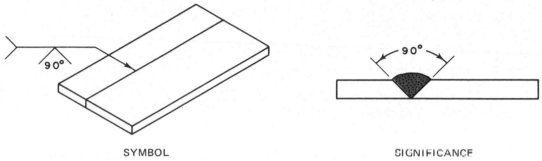

SYMBOL   SIGNIFICANCE

Fig. 35-1 Brazed beveled butt joint

MATERIALS

Two pieces of cast iron 1/4 inch or thicker, 2 in. X 6 in. each
1/8-inch diameter bronze rod
Suitable flux

PROCEDURE

1. Grind or machine one edge of each of the two pieces of cast iron to 45 degrees. When the two pieces are aligned, the included angle of this opening should be 90 degrees, figure 35-1.

   Note: The included angle for brazing is normally made larger than that for welding the same size material. The reason is that in a bonding process such as brazing, the wider V presents more bonding surface for the brazing alloy; as a result a stronger joint is formed.

2. Align the pieces and preheat them by playing the flame over the joint until the work becomes dull red in color.

3. Sprinkle some Hi-Bond® flux along the joint and tack each end of the assembly.

4. Braze in a manner similar to that in unit 33. Be sure that good tinning action takes place along the sides of the V. Do not try to complete the braze in one pass, figure 35-2. Attempts to build up too great a thickness in one step usually result in poor tinning and a weak joint.

5. After inspecting the braze for appearance, cut the piece in two and place each piece in a vise with the brazed joint slightly above the vise jaws. Bend one piece toward the root of the braze and one toward the face of the braze, figure 35-3.

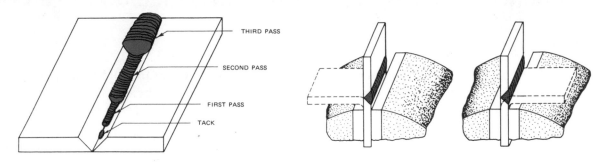

Fig. 35-2 Multilayer brazed joint   Fig. 35-3 Testing the braze

6. After these pieces break, inspect the break for a good bond. This is indicated by a coating of bronze on the cast iron and small particles of cast iron clinging to the bronze.

7. Prepare two more pieces. After beveling, draw a coarse file over the beveled surfaces to roughen them. This presents a surface of greater area to the brazing alloy. In addition, any free graphite which may be left on the beveled surface by the grinding operation is removed.

8. Braze this joint and test as before. Compare the bond strength of these plates with the first set tested.

REVIEW QUESTIONS

1. What prebrazing operation is most important to insure success in making a brazed joint?

2. What is the effect if brazing is attempted on a surface coated with free graphite?

3. When the joints made in steps 4 and 7 are tested, is there any difference in the amount of bending necessary to break the joints? Which joint requires the most bending? Why?

4. What color should the base metal have for good brazing?

5. Why is the included angle of the brazed bevel joint wider than for welding?

UNIT 36 SILVER SOLDERING NONFEROUS METALS

Silver soldering or *silver brazing* is, in reality, low-temperature brazing. A typical alloy, such as the one used in this unit, contains 80 percent copper, 15 percent silver, and 5 percent phosphorus. This type of alloy is effective at temperatures well below the melting point of brass and copper, the metals most commonly silver-soldered.

A number of different alloys with various melting points and fluxes is available.

The development of skill in silver soldering greatly increases the scope of the work which a welder can undertake.

MATERIALS

Two pieces of strip brass or copper 1/16-inch thick, 1 in. X 6 in.
Two pieces of brass or copper tubing, one of which just slips inside the other.
Handy-Flux® or equivalent
Sil-Fos® brazing alloy or equivalent, 1/16 in. X 1/8 in. X 14 in.
Airco® #3 welding tip or equivalent

PROCEDURE

1. Prepare the strips to be brazed by wire brushing, rubbing with emery cloth or steel wool, or by dipping in an acid bath to insure absolute cleanliness.

   Note: All welding and brazing operations are more successful if attention is given to surface cleanliness. In the silver brazing operations, failure to observe the proper precautions results in joints of low strength and poor appearance. A bright surface does not necessarily mean a clean surface from a welding or brazing viewpoint. Surface oxides sometimes appear very bright.

2. Apply a thin layer of flux to both surfaces of the strips that are to make contact in the braze. Allow this fluxed area to extend somewhat beyond the area to be brazed.

3. Set up the work so that the lapped surfaces to be brazed are in close contact, with the ends lapped 1 inch.

   Note: Proper joint spacing is very important in silver brazing and silver soldering. The ideal clearance between the surfaces to be brazed is 1 1/2 thousandths of an inch (.0015 inch). If this clearance is kept, the joint develops its maximum strength. The strength of the finished joint falls off rapidly as the clearance is increased beyond .0015 inch. At the same time, the amount of expensive silver soldering alloy used rises at a rapid rate.

4. Apply a 2X or 3X flame to the work with a back-and-forth motion to heat the brazing area. Do not apply the flame directly to the brazing rod.

5. Heat the work until the alloy can be applied to the joint a short distance from the flame. If the work is hot enough, the heat is conducted into the alloy causing it to melt and flow into the joint by capillary action.

6. During the brazing operation, the flux serves as a temperature guide, as well as a cleaning agent.

   Note: When the heat is first applied, the flux dries and turns white. As the amount of heat in the work increases, this white powder starts to melt, forming small beads of molten flux on the surface. Further heating causes the flux to become more fluid and flow out over the work surface in a thin, even coating. When this condition is noted, the work is at the proper brazing temperature, 1,300 degrees F. Temperatures beyond this point tend to make the molten flux *crawl* or leave bare areas on the work surfaces which oxidize. The result is a poor bond and unacceptable appearance.

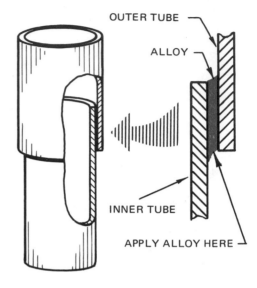

Fig. 36-1 Silver brazing tubes

7. Test the finished joint by trying to peel one of the lapped pieces from the other. If the braze has been properly made, the metal tears before the joint breaks.

8. Clean two pieces of copper or brass tubing inside and out. Make sure that one piece just slips inside the other, as in figure 36-1.

9. Place flux on both pieces and slide one into the other for a distance of 1 inch. Set this assembly on the bench or in a vise in a vertical position with the larger tube at the top.

10. Proceed with the brazing, making sure that all precautions outlined above are observed. Apply the alloy only on the outside of the joint. Make sure that the heating is complete.

11. Cool the finished work and note the thin, even line of alloy around the tube. Look into the tube and note that the alloy flowed by capillary action up between the two tubes and formed a small bead at the end of the inner tube. A further check can be made by hacksawing the joint diagonally and noting the white silvery appearance of the alloy between the entire lapped surfaces.

REVIEW QUESTIONS

1. If the base metal appears highly oxidized after the braze is completed, what factors must be checked before this condition can be corrected?

2. What are the factors that must be observed to produce good silver brazed joints?

3. What is a 3X flame?

4. What alloys are contained in typical silver solder?

5. How can material be prepared for silver soldering?

UNIT 37 SILVER SOLDERING FERROUS AND NONFERROUS METALS

Silver soldering, silver brazing, and low-temperature brazing are terms used for the same process in industry. To be technically correct, the process done at a temperature greater than 800 degrees F. is called *brazing*. The melting points for the different brazing alloys are:

Easy-Flo®	1,175° F.
Sil-Fos®	1,300° F.
Phos-Copper®	1,600° F.
Bronze	1,750° - 1,800° F.

The operations in this unit are done at a lower temperature than in the previous unit. The Easy-Flo® alloy is used with either ferrous or nonferrous metals.

All silver soldering or silver brazing operations require very small amounts of the alloy. Any attempt to use these alloys as welding rods or high-temperature brazing rods results in a weak and costly joint.

MATERIALS

Strips of steel, stainless steel, copper, and brass about 1/16-inch thick, 1 in. X 4 in. each
Handy-Flux®
Regular Easy-Flo® and #3 Easy-Flo® 1/16-inch diameter alloy
Airco® #3 welding tip

PROCEDURE

1. Prepare steel plates following the procedure of unit 36.

2. Set up two of the plates, lapping the ends about 1 inch to make an assembly 1 inch wide X 7 inches long. Be sure that the opposite end of the top plate is supported so that the two plates are kept parallel.

3. Braze the joint, using the same flame adjustment and technique as in unit 36. Apply the alloy (regular Easy-Flo®) by wiping the rod slowly on the one-inch face of the lap on the top side only. Apply enough heat to cause the alloy to flow between the lapped surfaces completely.

4. Cool the brazed joint and observe the opposite face of the lap and both edges. The alloy should be visible all around.

5. Try peeling one lapped plate from the other. This is impossible if the joint is properly made.

6. Set up two stainless steel plates and repeat the steps above.

   Note: The upper limit of temperature is very critical when brazing stainless steel. Overheating causes the formation of chromium oxide which can be removed only by filing or grinding. Flux does not remove chromium oxide.

7. Experiment with stainless steel joints by heating two of the plates until the color indicates that oxide has formed. Then proceed with the brazing as in previous joints and observe the results.

8. Check the joint of step 7 visually for complete bonding; peel one plate from the other.

   Note: Tensile strength tests made on properly brazed joints on stainless steel break outside the brazed area at a load of 100,000 to 120,000 pounds per square inch. The actual strength of the braze is somewhat higher than the above figures.

9. Set up two brass plates to make a fillet- or tee-type joint. Make sure both sides of the joint are covered with flux.

10. Braze this joint using #3 Easy-Flo®. Note that this alloy does not flow as freely as other types, but allows a slight fillet to build up. This is desirable in certain applications.

11. Cool and inspect the finished braze. In particular, check the draw through of the alloy on the reverse side of the joint.

12. Hammer the upstanding leg flat against the bottom plate. Note that the color of the alloy is very close to the color of the work. This is desirable in many jobs where color match is important.

13. Prepare two more plates but, in this case, make a butt joint. Test in the same manner as for a butt weld and observe the results.

REVIEW QUESTIONS

1. If the joint is overheated when low-temperature brazing stainless steel, what procedure must be followed to obtain a satifactory braze?

2. Is it possible to make fillets when using silver brazing alloys? Explain.

3. What is the proper clearance between the lapping surfaces for maximum joint strength?

4. How do butt-brazed joints compare with lapped joints?

5. What can be said about the amount of filler metal required for a given joint?

SECTION 2
Arc Welding

In electric arc welding, a high temperature arc melts the metals to be joined while additional molten metal is being added. After the metal has solidified, the resulting joint will be stronger than the original metal.

This type of welding became a recognized industrial process in the early part of this century. The development of arc welding paralleled the development of electrical apparatus and sources of electricity.

Arc welding is widely used in the manfacturing and construction industries. Special applications of basic arc welding such as resurfacing steel parts, hard facing, automatic welding processes, pipe welding and shielded arc welding are widely used.

Many different occupations require knowledge and skill in arc welding. Due to the variables encountered and the complex equipment used, the arc welder must possess a high degree of skill and considerable technical knowledge.

UNIT 38   THE ARC WELDING PROCESS

One of the most important processes in industry is the fusion of metals by an electric arc. This is commonly called *arc welding.*

Briefly, the process takes place in the following manner. The work to be welded is connected to one side of an electric circuit, and a metal electrode is connected to the other side. These two parts of the circuit are brought together and then separated slightly. The electric current jumps the gap and causes a continuous spark called an *arc.* The high temperature of this arc melts the metal to be welded, forming a molten puddle. The electrode also melts and adds metal to the puddle, figure 38-1.

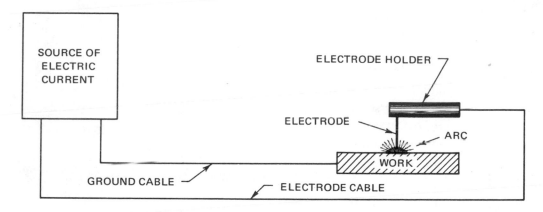

Fig. 38-1 Simple welding circuit

As the arc is moved, the metal solidifies. The metal fuses into one piece as it solidifies.

The melting action is controlled by changing the amount of electric current which flows across the arc and by changing the size of the electrode.

HAZARDS

Before arc welding is begun, the student should be fully aware of the personal dangers involved. Of course, the high-temperature arc and the hot metal can cause severe burns. However, the electric arc itself may be a hazard.

An electric arc gives off large amounts of ultraviolet and infrared rays. Infrared rays are also given off from the molten metal. Both types of rays given off from arc welding are invisible, just as they are when given off from the sun. They will cause sunburn, the same as they will from the sun, except that the rays given off from the electric arc, burn much more rapidly and deeply. Since these rays are produced very close to the operator, they can cause severe damage to the eyes in a very short period of time.

During arc welding there is a danger that small droplets of molten metal may leave the arc and fly in all directions. These so-called sparks range in temperature from 2,000 degrees F.

(1093 degrees C) to 3,000 degrees F. (1649 degrees C) and in size from very small to as large as 1/4 inch in diameter. They may cause burns plus they are a fire hazard when they fall on flammable material.

PROTECTIVE DEVICES

For protection from the rays of the arc and the flying sparks, the welding operator must use a helmet, figure 38-2, and other protective devices. The welding helmet is fitted with filter plates that screen out over 99% of the harmful rays. The helmet must be in place before attempting to do any arc welding. The arc is harmful up to a distance of 50 feet and all persons within this range must be careful that the rays do not reach their eyes.

Most welding helmets are made of pressed composition material or molded plastic. If they are dropped or if material is dropped on them they may be unfit for use. Each helmet is equipped with an adjustable headband. Any attempt to use wrenches or pliers to force the adjusting device may destroy the helmet.

The filter plate in each helmet is a special, costly glass which should be handled with great care. All filter plates should be protected from the flying globules of molten metal in the manner indicated in figure 38-3.

All welding stations should be equipped with curtains or other devices which keep the arc rays confined to the welding area. For the protection of others, the welder should make sure that these curtains are in place before starting any welding.

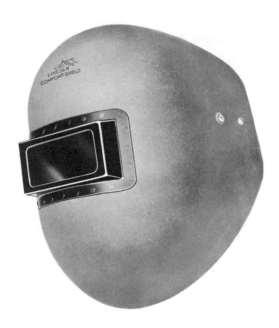

Fig. 38-2 A modern welding helmet

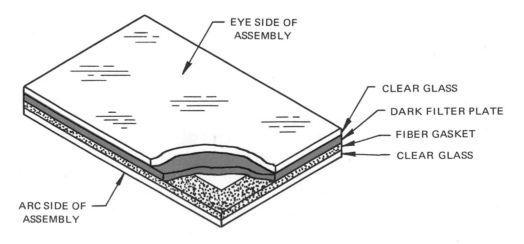

EYE SIDE OF ASSEMBLY

CLEAR GLASS

DARK FILTER PLATE

FIBER GASKET

CLEAR GLASS

ARC SIDE OF ASSEMBLY

Fig. 38-3 Filter plate protective assembly

Fig. 38-4 Note protective devices

The arc rays will penetrate one thickness of cloth and cause sunburn. Therefore, the operator must protect himself with fire-resistant aprons, sleeves, and gloves to eliminate this hazard plus the hazard of fire, figure 38-4. In fact, all clothing worn by the operator should be reasonably flame resistant. Clothing which has a fuzzy surface can be a serious fire hazard, particularly if it is cotton.

Naturally, all other types of flammable material such as oil, wood, paper, and waste should be removed from the welding area before any welding is attempted. Each welder should be acquainted with the location and operating characteristics of all fire extinguishers.

REVIEW QUESTIONS

1. Why is clear glass used on the eye side of the filter plate assembly in figure 38-3?

2. What determines how fast the weld metal melts?

3. Who is responsible for the protection of workers in the area of the welding operation?

4. What type of rays are given off by the electric arc?

5. Can the rays given off by the electric arc be seen?

UNIT 39 SOURCES OF ELECTRICITY FOR WELDING

TYPES OF WELDING MACHINES

Electric current for the welding arc is generally provided by one of two methods. A transformer which reduces the line voltage can provide *alternating current* (AC). This current reverses direction 120 times per second. The transformer has no moving parts.

Direct current (DC) for the welding arc may be produced by a direct-current generator connected by a shaft to an AC motor, figure 39-1. A gasoline engine or other type of power may also be used to turn the generator. Direct current flows in the same direction at all times. In any case, the welding machine must have the ability to respond to the need for rapid changes in the welding voltage and current.

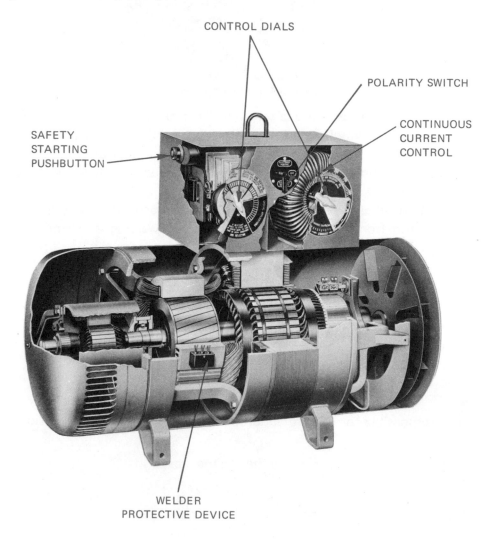

CONTROL DIALS

POLARITY SWITCH

CONTINUOUS
CURRENT
CONTROL

SAFETY
STARTING
PUSHBUTTON

WELDER
PROTECTIVE DEVICE

Fig. 39-1 A motor generator-type of DC welding machine

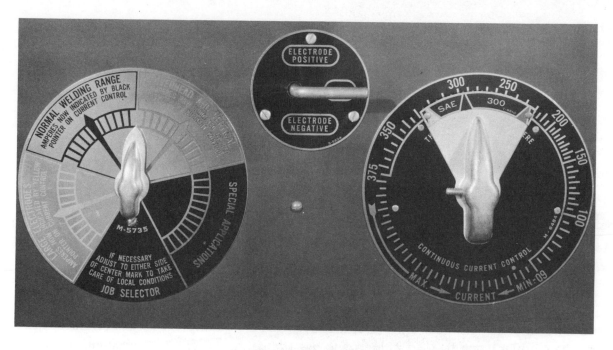

Fig. 39-2 Dual-current control

Both types of machines are widely used in industry, but the DC type is slightly more popular. Both are supplied in various sizes, depending on the use to which they are to be put, and are designated by the maximum continuous current in amperes which they can supply, (e.g., 150a or 300a).

CURRENT CONTROL

Different types of welding operations require different amounts of current (amperes). Therefore, arc welding machines must have a way of changing the amount of current flowing to the arc.

The DC generator may have either a *dual-current control,* figure 39-2, (the more popular type) or a *single-current control.* In the dual-current control type, two hand-wheels or knobs adjust the electrical circuits to provide the proper current to the arc. In the single-control

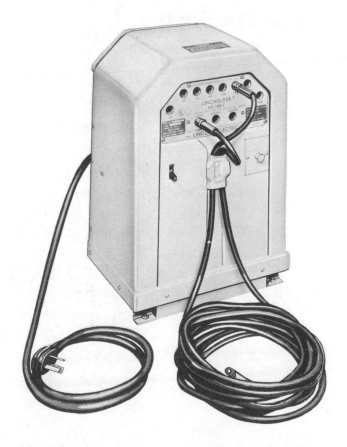

Fig. 39-3 AC machine with taps for current control

type, a single wheel adjusts the current. The other controls on a DC machine are an on-off switch and a polarity switch.

In one type of current control for the AC machine, the output cables are moved from one tap to another on the secondary side of the transformer winding, figure 39-3. The disadvantage of this type is that the number of current combinations is limited by the number of taps in the machine.

The other popular type of AC machine has a movable core in the transformer. On this type of machine the operator selects the desired current rating by turning a handwheel.

Another type of machine which has gained popularity is the transformer-type with a built-in rectifier. The rectifier converts alternating current to direct

Fig. 39-4 AC/DC machine — single-current control

current for welding. In some types of machines, the alternating current can be taken ahead of the rectifier when it is advantageous to use alternating current in the welding operation, figure 39-4.

CARE AND PRECAUTIONS

- When a motor-generator type of welding machine is turned on, the operator should immediately check to see if the armature is rotating and that the direction of rotation is correct according to the arrow on the unit.

- Occasionally a fuse blows or a starter contact becomes burned or worn. Either of these conditions may cause the machine to overheat if it is left on. This can rapidly damage the welding machine.

- Starter boxes and fuse boxes carrying 220 or 440 volts should not be opened by the operator.

- The welding cable terminal lugs should be clean and securely fastened to the terminal posts of the machine. Loose or dirty electrical connections tend to overheat and cause damage to the terminal posts.

REFERENCE

Manufacturer's bulletin for the machine to be used.

REVIEW QUESTIONS

1. What is a direct-current welding circuit?

2. What does the term current control mean?

3. What effect do loose connections have on the welding circuit?

4. What are the three types of welding machines?

5. How many times per second does alternating current reverse direction?

UNIT 40   THE WELDING CIRCUIT

PARTS OF THE WELDING CIRCUIT

In addition to the source of current, the welding circuit consists of:

- The work
- The welding cables
- The electrode holder
- The electrode

The *work* with which the welder is concerned may be steel plate, pipe, and structural shapes of varying sizes and thicknesses. It should be suitably positioned for the job being done. The work is a conductor of electricity and thus, is a part of the circuit.

The *welding cables,* figure 40-1, are flexible, rubber-covered copper cables of a large enough size to carry the necessary current to the work and to the electrode holder without overheating. The size of the cable depends on the capacity of the machine, and the distance from the work to the machine. A *ground clamp,* figure 40-2, is attached to the end of one of the cables, so that it may be connected to the work.

The *electrode holder,* figure 40-3, is a mechanical device on the end of the welding cable which clamps the welding rod or electrode in the desired position. It also provides

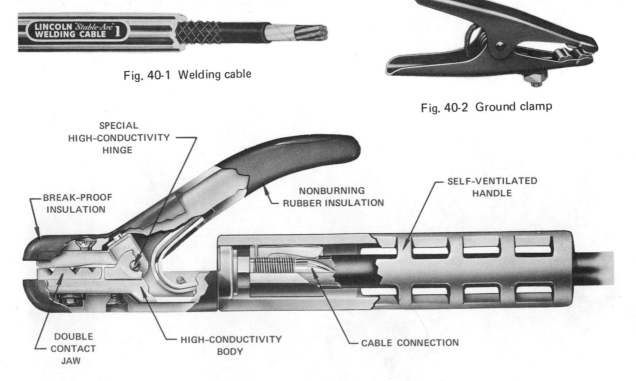

Fig. 40-1 Welding cable

Fig. 40-2 Ground clamp

Fig. 40-3 A cutaway view of an electrode holder

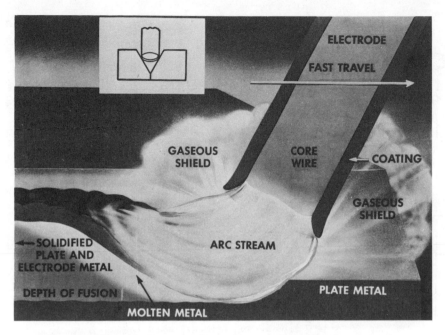

Fig. 40-4 Coated electrodes form a gaseous shield around the arc

an insulated handle, with which the operator can direct the electrode and arc. These holders come in various sizes depending on the amperage which they are required to carry to the electrode.

THE ELECTRODE

The electrode usually has a steel core. This core is covered with a coating containing several elements, some of which burn under the heat of the arc to form a gaseous shield around the arc, figure 40-4. This shield keeps the harmful oxygen and nitrogen in the atmosphere away from the welding area.

Other elements in the coating melt and form a protective slag over the finished weld. This slag promotes slower cooling and also protects the finished weld or bead from the atmosphere. Some coated electrodes are designed with alloying elements in the coating which change the chemical and physical characteristics of the deposited weld metal.

The result of using properly designed coated electrodes is a weld metal which has the same characteristics as the work, or base metal, being joined.

Electrodes are supplied commercially in a variety of lengths and diameters. In addition, they are supplied in a wide variety of coatings for specific job applications. These applications are discussed in other units.

The American Welding Society and The National Electrical Manufacturer's Association classify electrodes according to the type of coating, operating characteristics, and chemical composition of the weld metal produced. Chart 40-1 indicates some of these electrodes with their color code markings. These represent the commonly used rods. There are many more. Most electrode manufacturers supply, free of charge, a chart of all the electrodes they make.

A growing number of electrode manufacturers mark the grip end of the electrodes with their classification, either EXXXX, as indicated in Chart 40-1, or simply XXXX so the welder does not have to memorize the color code.

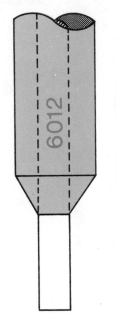

AWS NUMBER LOCATION

COLOR CODE LOCATION

AWS Classification	Type	Current and Polarity	Tensile Strength P.S.I.	Yield Point P.S.I.	% Elongation in 2 inches	Color Code			Applications
						End	Spot	Group	
E-4510	Bare	d.c. Negative	45-50,000	40-45,000	5-7%	None	None	None	Building up worn surfaces and training operators.
E-6010	Coated	d.c. Positive	65-72,000	53-58,000	27-30%	None	None	None	All-position, all-purpose high-impact, high-ductility code welding
E-6011	Coated	a.c. or d.c. Positive	65-72,000	53-58,000	27-30%	None	Blue	None	All-position, primarily for producing welds with a.c. current equal to E-6010
E-6012	Coated	a.c. or d.c. Negative	68-78,000	58-68,000	20%	None	White	None	All-position, for poor "fitup" work and work where resistance to impact and low ductility are not too important.
E-6013	Coated	a.c. or d.c. Negative	75,000	62,000	20%	None	Brown	None	For all-position work primarily with a.c. current. Compares to E-6012 series in general applications
E-6020	Coated	a.c. or d.c. Negative	68,000	56,000	32%	None	Green	None	For flat-position welding, used in code welding. Being replaced slowly by E-6027 and E-7024
E-7014	Coated Iron Powder	a.c. or d.c. Negative	72-82,000	62-72,000	23-32%	Black	Brown	None	High-speed production work, faster deposition rate than E-6012 or E-6013
E-7024	Coated Heavy Iron Powder	a.c. or d.c.	75-83,000	63-75,000	17-25%	Black	Yellow	None	Fast deposition rate, excellent appearance – for joints not requiring deep penetration
E-6027	Coated Iron Powder	a.c. or d.c. Negative	62-69,000	52-60,000	25-35%	None	Silver	None	Iron powder version of E-6020 – faster deposition for flat and horizontal positions

Chart 40-1 Mild Steel Electrode Chart

Only the code markings are important when determining the type of electrode. The coating colors should not be depended on for recognition as they vary with manufacturers. Some producers also place dots and other markings on the coating. These are only trademarks, and are not to be confused with the code markings, which appear only on the grip end of the electrodes.

Any of the electrodes shown in Chart 40-1 may be purchased in a wide variety of sizes and lengths.

AWS CLASSIFICATION NUMBERS

All AWS (American Welding Society) numbers consist of three parts. For example, E-6010.

1. The E in all cases indicates an electric arc welding electrode or rod.

2. The number following the E (in this case, 60) indicates the minimum tensile strength of the weld metal in thousands of pounds per square inch, (in this case, 60,000 p.s.i.). This number could be 80, 100, or 120 which would indicate minimum tensile strengths respectively of 80,000 p.s.i., 100,000 p.s.i., or 120,000 p.s.i.

3. In a four-digit number the third digit indicates the positions in which the electrode may be used: 1—indicates all positions; 2—flat or horizontal; 3—deep groove.

4. The fourth digit indicates the operating characteristics, such as polarity, type of covering, bead contour, etc.

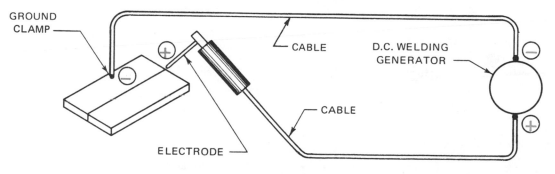

POSITIVE OR REVERSE POLARITY

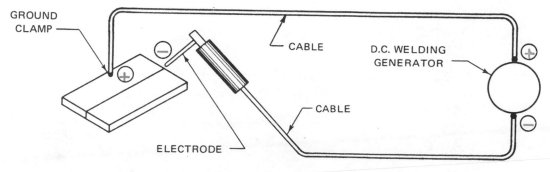

NEGATIVE OR STRAIGHT POLARITY

Fig. 40-5 The welding circuit

POLARITY

In welding with direct current, the electrode must be connected to the correct terminal of the welding machine. This polarity may be changed by a switch on the welding machine. The polarity to be used is determined by the type of electrode and is indicated in the electrode chart, Chart 40-1.

When the electrode is connected to the negative terminal (−), the polarity is called *negative or straight.* When connected to the positive terminal (+), it is called *positive or reverse,* figure 40-5. The use of incorrect polarity produces a poor weld. When welding with alternating current, polarity is not considered.

A simple test for checking the polarity of an electric welding machine is as follows:

1. Place a carbon electrode in the electrode holder.

2. Strike an arc. Maintain the puddle and weld for 5 or 6 inches.

3. Check the plate for smears or black smudges. If these are present, the machine is in reverse polarity.

REFERENCE

Manufacturer's chart of electrodes

REVIEW QUESTIONS

1. What is the primary purpose of the AWS code markings on welding rods?

2. What are the effects of oxides and nitrides in the weld metal?

3. Referring to Chart 40-1, what AWS type of electrode is used if the strength of the weld is the only characteristic of importance for a given job?

4. If it is necessary to weld pressure pipe in the overhead position and the only machine available is an alternating-current type (AC), what AWS classification rod is used?

UNIT 41   FUNDAMENTALS OF ARC WELDING

VARIABLES

Four things greatly affect the results obtained in electric arc welding. To make good welds, each one must be adjusted to fit the type of work done and the equipment being used.

They are:

- Current setting or amperage

- Length of arc or arc voltage

- Rate of travel

- Angle of the electrode

CURRENT SETTING

The current which the welding machine supplies to the arc must change with the size of the electrode being used. Large electrodes use more current than smaller sizes. A good general rule to follow is: when welding with standard coated electrodes, the current setting should be equal to the diameter of the electrode in thousandths of an inch.

Thus, a 1/8-inch electrode measures .125 inch and operates well at 125 $\pm$ a few amperes. Similarly, a 5/32-inch rod measures .156 inch and operates well at 150 $\pm$ a few amperes. The $\pm$ indicates that these electrodes will operate well in a range of current values either below or above the indicated amperage. For example, a value of 125 $\pm$ 10 amperes indicates a range of values with a low of 115 amperes and a high of 135 amperes.

When indicating the diameter of the electrodes, reference is made only to the steel or alloy core of the rod, figure 41-1. The overall diameter including the rod coating is not the indicated electrode size.

Fig. 41-1 Measure core of rod

LENGTH OF ARC

The arc length is one of the most important considerations in arc welding. Variations in arc length produce varying results.

The arc length increases as the arc voltage increases. For example, an arc 3/16 inch long requires three times the voltage of a 1/16-inch arc, figure 41-2.

The general rule on arc length states: The arc length shall be slightly less than the diameter of the electrode being used.

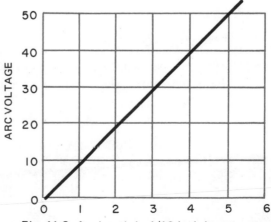

Fig. 41-2 Arc length in 1/16-inch increments

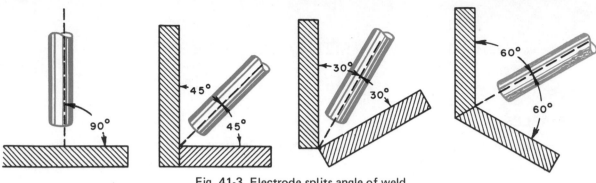

Fig. 41-3 Electrode splits angle of weld

Thus, a 5/32-inch diameter electrode operates well between 1/8 inch and 5/32 inch of arc gap, or 20-22 arc volts according to the chart, figure 41-2.

It is almost impossible for the operator to measure the arc length accurately when welding. However, the welder can be guided by the sound of the arc. At the proper arc length, the sound is a sharp, energetic crackle. Proper arc length is determined by noting the difference in the sound of the arc when it is set too far, and at just about the right distance from the work. By practicing this, the operator will be able to judge good arc length by the distinctive sound.

RATE OF TRAVEL

The *rate of travel* of the arc changes with the thickness of the metal being welded, the amount of current, and the size and shape of the weld, or bead, desired.

The welding student should begin by making welds known as single-pass stringer beads. The arc length and arc travel should be such that the puddle of molten metal is about twice the diameter of the rod used.

ANGLE OF ELECTRODES

When welding on plates in a flat position, the electrode should make an angle of 90 degrees with the work. In other than flat work, good results are obtained if the rod splits whatever angle is being welded, figure 41-3. In general practice it is found that this angle may vary as much as 15 degrees in any direction without affecting the appearance and quality of the weld. The electrode angle should be no greater than 20 degrees toward the direction of travel.

REVIEW QUESTIONS

1. Using the general rule for current setting, what is the proper setting, to the nearest round figure for electrodes with the following diameters: 3/32 inch, 3/16 inch, and 1/4 inch?

2. From figure 41-2 and the general rule for arc length, what is the voltage across the arc when welding with a 3/16-inch electrode?

3. What is the arc voltage for a 1/8-inch diameter electrode?

4. If the first attempt at making a stringer bead produces a weld that is too narrow, what adjustment must be made in the rate of travel to produce a bead of the proper width?

5. What is the proper angle of the electrode in the following sketch?

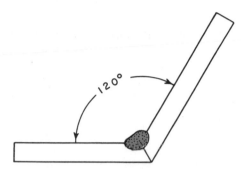

6. What indication does the operator have that the arc is the correct length for the diameter rod being used?

UNIT 42   STARTING AN ARC AND
RUNNING STRINGER BEADS

The quality and appearance of an electric arc weld depend almost entirely on the following:

- Length of the arc

- Rate of Travel

- Angle of the electrode

- Amount of current

Experimentation with each of these variables is helpful in learning correct welding procedures. This unit provides an opportunity to experiment with these variables and observe the results.

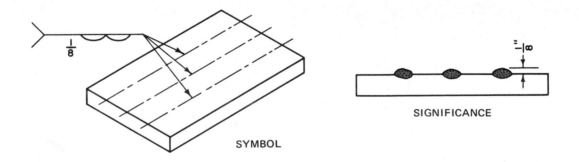

SYMBOL

SIGNIFICANCE

Fig. 42-1 Stringer beads

Materials

Steel plate 3/16″ or thicker 6 in. x 9 to 12 in.

DC or AC welding machine

1/8-inch or 5/32-inch diameter E-6012 or E-6013 electrodes

Procedure

1. Start the machine, check the polarity and adjust the current setting as described in unit 41.

   CAUTION: Make sure all protective devices are in place. Use all recommended safety devices to protect the body, especially the eyes, from the arc rays. Failure to do so results in severe and painful radiation burns. Wear safety glasses.

2. Start the arc on the plate according to the method indicated in figure 42-2.

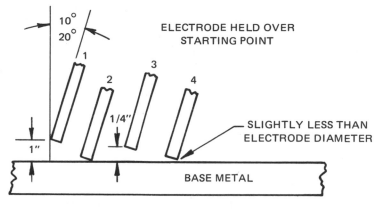

1. HOLD ELECTRODE 1" OFF PLATE, BRING HELMET OVER EYES
2. TOUCH PLATE WITH ELECTRODE
3. RAISE ELECTRODE 1/4"
4. LOWER TO NORMAL ARC LENGTH (UNIT 4)

10°
20°

ELECTRODE HELD OVER STARTING POINT

1   2   3   4

1/4"

SLIGHTLY LESS THAN ELECTRODE DIAMETER

1"

BASE METAL

Fig. 43-2 Establishing an arc

3. Listen for the sound indicating the correct arc length, and observe the behavior of the arc.

4. Make straight beads or welds. Note that the electrode must be fed downward at a constant rate to keep the right arc length. Move the arc forward at a constant rate to form the bead.

   Note: Right-handed welders will see better welding from left to right.
   Left-handed welders should weld from right to left.

5. Remove the slag and examine the bead for uniformity of height and width.

   CAUTION: When removing slag from a weld with a chipping hammer, eye protection is very important. Safety glasses should always be worn.

6. Continue to make stringer beads until each weld is smooth and uniform.

7. Make a series of beads similar to those in figure 42-3. Note the difference of each bead as the variables are changed.

A. NORMAL BEAD
B. ARC TOO LONG
C. ARC TOO SHORT
D. RATE OF TRAVEL TOO HIGH
E. RATE OF TRAVEL TOO LOW
F. CURRENT TOO LOW
G. CURRENT TOO HIGH
H. ROD ANGLE TOO LOW

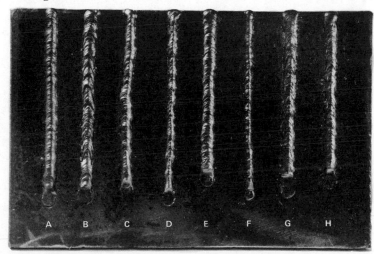

Fig. 42-3 Bead variables using E-6012 electrode and DC straight polarity.

AWS-ASTM Class	Hobart	Air Products	Airco	Canadian Liquid Air Ltd.	Canadian Rockwell	P & H	Lincoln	Marquette	McKay Co.	M & T (Murex)	N.C.G. (Sureweld)	Reid-Avery Co. (Raco)	Shober	Canadian Liquid Carbonic	Westinghouse
E-4510 E-4520	Sulkote		41 63	LA-SC-15		Washcoat	Stable Arc	101	21 3	Sulcoat Thincoat	4510	Raco Type D & M			Sulcoat 18
E-6010	10 10-IP	6010 6010-IP	6010	LA-6010 6010-P	R60	SW-610 AP-100	Fleetweld 5 Fleetweld 5P	105	6010 6010-IP	Speedex 610	6010-Y 6010-X	6010	32	P&H-704D 610-P	XL-610 XL-610A ZIP-10
E-6011	335-A	6011 6011-C	6011 6011-C	LA-6011-P 6011-F	R61	SW-14 SW-14-IMP	Fleetweld 35, 180 35-LS	130	6011 6011-IP	Type A 611-C	6011-Z	6011 6011-IP	11 13	P&H-504D 611-P	ACP-611
E-6012	12 212-A 12-A	6012-GP 6012-SF 6012-IP	6012 6012-C	LA-6012 6012-P	R62	SW-612 PFA SW-17 SW-29	Fleetweld 7, 7MP 77	120	6012	Genex-M Type N-13	6012W	6012 6012-F	34	604	FP-612 FP-2-612 ZIP-12 ZIP-AF
E-6013	13A 413 447-A	6013-GP 6013-SF	6013 6013-C	LA-6013 6013-P	R63	AC-3 SW-16 SW-15	Fleetweld 37	140	6013	Murex U, U-13	6013-Y	6013	13 35	613	SW-613 SW-2M-613
E-7014	714 14-A	7014-IP	Easyarc 7014	LA-7014	R74	DH-6 SW-15-IP	Fleetweld 47	146	7014	Speedex U		7014	14	P&H-714	ZIP-14
E-6020	111	6020	6020	LA-6020FS	R620	DH-3			6020	Murex FHP,D	6020-X				DH-620
E-7024	24 24-A	7024-IP	7024	6024	R724	SW-44 SW-624	Jetweld 1 3	24	7024	Speedex 24			36	P&H-624	ZIP-24
E-6027	27	6027-IP	6027	LA-6027	R627	DH-27	Jetweld 2			Speedex 27				P&H-727	ZIP-27
E-7010-A1	710	7010-A1	7010-A1	LA-7010-P		SW-75	Shieldarc 85			710-Mo		7010-A1		P&H-710	
E-7020-A1	111-HT	7020-A1	7020-A1				Jetweld 2HT			Murex DM		7020-A1			

Chart 42-1 Comparative index of mild steel and alloy electrodes

REVIEW QUESTIONS

1. What must be controlled to make good arc welds?

2. What is a stringer bead?

3. What are the points to look for in a good weld?

4. What is the current value for a 5/32-inch diameter electrode?

5. What two factors best determine a correct arc?

UNIT 43 RUNNING CONTINUOUS STRINGER BEADS

Running long stringer beads demands good control of the welding electrode if the beads are to be straight and uniform in appearance and size. Much practice is required to develop a high degree of skill.

Changing electrodes in the middle of the bead, or starting an arc which has been accidentally stopped is a basic and important skill. This unit provides experience in restarting a bead.

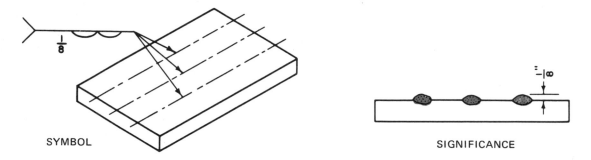

SYMBOL

SIGNIFICANCE

Fig. 43-1 Stringer beads

Materials

Steel plate, 3/16 inch thick, 6 in. x 9 to 12 in.

DC or AC welding machine

1/8- or 5/32-inch diameter E-6012 or E-6013 electrodes

Procedure

1. Start the machine, check the polarity, and adjust the current for the size of electrode being used.

2. Run a continuous stringer bead on the plate, using the full length of the electrode before stopping. Make this bead parallel to and about 1/2 inch from the edge of the plate.

3. Run additional beads at 1/2-inch intervals, being sure to keep each bead straight. Check the arc length and rate of travel constantly to produce smooth, uniform beads.

4. Continue to make this type of weld until each one is of uniform appearance for its entire length.

5. Make a bead 2 or 3 inches long and stop the arc. Start the arc again ahead of the crater. Move the electrode back to the crater, using an extra long arc. Bring the rod down rapidly to the proper arc length and make sure that the new puddle just fuses into the last ripple of the crater. Proceed with the weld for another 2 or 3 inches and stop.

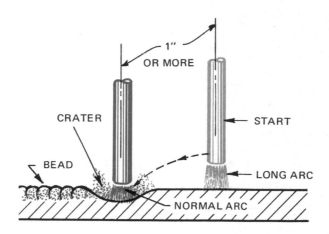

Fig. 43-2 Establishing a continuous bead

6. Continue this procedure until there is very little difference in appearance at the point where the arc was restarted. Figure 43-2 shows the right procedure.

7. Start an arc directly in the crater and notice the difference in the appearance of the connection.

   Note: The ideal length at which electrode stubs should be discarded is 1-1/2".

8. Try to eliminate the crater at the end of the finished bead by moving the arc back over the crater and finished bead slowly. As the arc is moved back gradually increase the arc length until the crater is filled as in figure 43-3.

9. Try to eliminate the crater by using a very short arc in the crater and pausing until there is enough buildup.

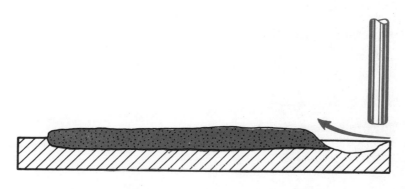

Fig. 43-3 Rod motion for filling crater

REVIEW QUESTIONS

1. What is the difficulty in trying to restart an arc directly in the crater?

2. What are the advantages of restarting a long arc ahead of the crater then backing up to the crater and shortening the arc?

3. Why is it necessary to gain skill in connecting the beads?

4. Why is it important for a welding student to learn to run a bead in a straight line?

5. What does the term polarity mean?

UNIT 44   RUNNING WEAVE BEADS

It is often necessary when welding large joints or making cover passes to produce beads wider than stringer beads. These are called *weave beads*. Weaving is done with a back and forth sidewise motion of the electrode and a slow forward movement, figure 44-1.

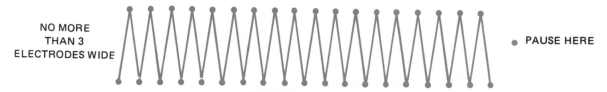

NO MORE THAN 3 ELECTRODES WIDE

PAUSE HERE

Fig. 44-1 Straight weave

The height of the bead depends on the amount the electrode is advanced from one weave to the next. The number of ripples depends on the speed and frequency of the weaving motions.

The pause at each side of the weave is important for puddle flow and penetration. Failure to pause causes an undercut along the sides of the weld, figure 44-2.

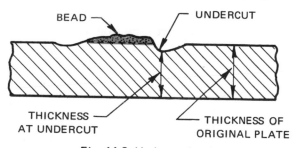

BEAD — UNDERCUT

THICKNESS AT UNDERCUT

THICKNESS OF ORIGINAL PLATE

Fig. 44-2 Undercut bead

Materials

Steel plate

1/8- or 5/32-inch diameter E-6012 or E-6013 electrodes

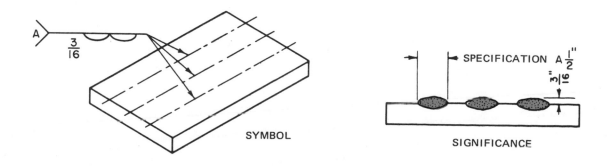

$\frac{3}{16}$

SYMBOL

SPECIFICATION A $1\frac{1}{2}$"

$\frac{3}{16}$"

SIGNIFICANCE

Fig. 44-3 Weave beads

Procedure

1. Start the machine, check the polarity, and adjust the current for the size of the electrode being used.

2. Start an arc and make a bead approximately 3 times wider than the diameter of the electrode being used.

3. Continue to make weave beads until they have a uniform height and width for their entire length.

   Note: Beginners have a tendency to let the weld become progressively wider with each pass of the arc. In an attempt to correct this, there is a tendency to decrease each weave motion.

4. Try stopping the arc and fusing the new bead to the original. Be sure to *slag* (remove the slag with a wire brush or hammer) the crater before starting each bead. Practice this until the starting and stopping points blend in smoothly.

5. Change the length of pause at each side. Change the travel speed and the amount of advance.

6. Clean the beads and compare the results.

REVIEW QUESTIONS

1. Why should there be a definite pause at each side of the weave?

2. What can be done to produce a weave bead with many fine ripples rather than a few coarse ones?

3. When restarting the bead, how does the rate of travel for the first two or three passes compare with the normal rate of travel?

4. What steps are taken to prevent undercutting?

5. How should the size of the molten puddle compare with the diameter of the electrode?

UNIT 45 PADDING A PLATE

Experience in padding helps the welding student develop an eye for following a joint It helps the student compare beads for uniform appearance. Padding is used for building up pieces prior to machining and for depositing hardfacing metal on construction equipment.

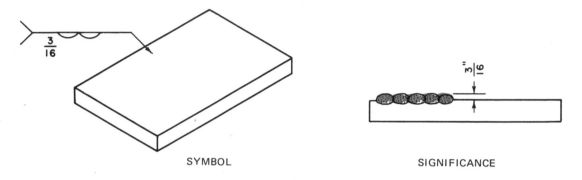

SYMBOL   SIGNIFICANCE

Fig. 45-1 The pad

Materials

2 Pieces of plate 6 in. x 6 in.

1/8- or 5/32-inch E-6012 or E-6013 electrodes

Procedure

1. Establish an arc and run a stringer bead close to and parallel with the far edge of the practice plate.

   Note: Observe that a crater is left at the end of this weld. Prevent this crater in the following manner: Upon reaching the end of the plate, pull the electrode out of the crater, letting the heat die down. When the color has disappeared restart the arc in the crater, depositing a small amount of weld. This can be done several times to fill the crater.

2. Run more beads alongside of the previous bead, figure 45-2. Make sure that the far edge of bead being deposited is in the center of the previous bead.

Fig. 45-2 A partially completed pad using E-6012 electrode (right-handed operator welding left to right).

Note: The electrode must be directed at the point where the previous bead meets the base metal.

Always chip the slag from each bead before welding.

3. Continue to cover the plate with weld using this technique. Keep the weld as straight as possible.

4. On another plate follow the same procedure and make a pad using the weave technique.

Note: As the newly padded surface cools, the plate may bow upward. This can be corrected by welding a pad on the opposite side of the plate.

5. For additional practice, the plate can be turned 90 degrees for a second layer.

6. After the padding operation is finished, cut the material by saw or torch and examine the weld. There should be no holes or bits of slag imbedded in the weld.

REVIEW QUESTIONS

1. What is weld padding?

2. What characteristics of a weld are most easily examined through padding?

3. Where can the padding operation be used?

4. Why is it important to slag every bead before running the next one?

5. How is a washed-out area at the edge of a plate prevented?

UNIT 46 SINGLE-PASS, CLOSED SQUARE BUTT JOINT

In making a single-pass, closed square butt joint, penetration is extremely important. Welding from one side not only makes complete penetration difficult, but the joint strength depends directly on the depth of penetration.

By experimenting with this type of joint and testing the results, one can determine the best procedures.

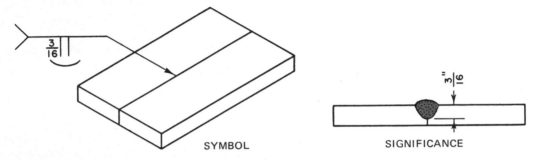

SYMBOL   SIGNIFICANCE

Fig. 46-1 Single-pass, closed square butt joint

Materials

Two steel plates, 1/4 inch thick, 2 in. x 9 in. each

1/8- or 5/32-inch diameter E-6012 or E-6013 electrodes

Procedure

1. Place the plates on the worktable so that the two 9-inch edges are in close contact.

2. Tack the two plates together using *tack welds* about 1/2 inch long. Start the tacks 1/2 inch to 1 inch from the ends of the plate to avoid having excessive metal and poor penetration at the start of the weld.

3. Proceed with the weld as in making stringer beads, but be very careful to keep the centerline of the arc exactly centered on the joint. Half the weld should be deposited on each plate.

4. Cool the finished assembly. Clean the bead and examine it for uniformity.

5. Check the depth of penetration of the weld by placing the assembly in a vise with the center of the weld slightly above and parallel to the jaws. Bend the plate toward the face of the weld so that the joint opens, figure 46-2. Examine the original plate edges.

6. Continue the bend until the weld breaks. Notice that the broken weld metal has a bright, shiny appearance, and that the metal that was not welded is much darker. This bright weld metal indicates the depth of penetration.

7. Make more joints of this type. Start with a setting of 150 amps and increase the amperage with each joint.

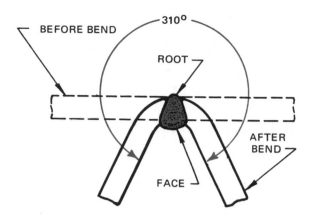

Fig. 46-2 Bend test for butt weld

8. Cool, break, and examine these test plates and compare the amount of penetration with that in the first weld made.

9. Set up another test plate and weld, using a weaving figure-8 motion, figure 46-3.

10. Cool, break, and examine this plate and compare the penetration with the other welds that have been made.

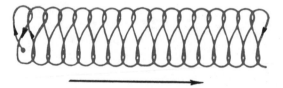

Fig. 46-3 Figure-8 weave

REVIEW QUESTIONS

1. How does increasing the amperage affect the depth of penetration?

2. How does the figure-8 weave bead affect the depth of penetration? Why?

3. Is there any advantage if the figure-8 weave is used in buildup operations?

4. What is penetration?

5. What is the purpose of a tack weld?

UNIT 47 OPEN SQUARE BUTT JOINT

An open butt joint presents some additional problems in penetration. By changing the space between the plates, and by welding one set of plates from one side only and another set from both sides, it is possible to compare the quality of the welds and, particularly, the penetration. The open square butt joint differs from the closed butt joint in that the open-type joint has some spacing between the plate edges.

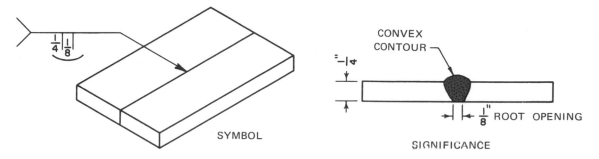

SYMBOL

SIGNIFICANCE

Fig. 47-1 Open square butt joint

Materials

Two steel plates, 1/4 inch thick, 1-1/2 to 2 in. x 9 in. each

5/32-inch diameter E-6012 electrodes

Procedure

1. Place the two plates on the welding bench, align and space them, and tack them as shown in figure 47-2. Tacks should be long enough to withstand the strain of the expanding metal being welded without cracking.

2. Make the weld in the same manner as a closed butt joint, but use a slight amount of weaving to allow for the additional width of the joint.

3. Cool, clean, and inspect the finished weld for uniform appearance. Examine the root side of the weld for penetration.

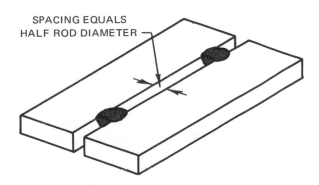

Fig. 47-2 Setup for open butt joint

4. Break the welded joint in the same manner as the closed butt joint. Check the amount of penetration. It should be a little more than one-half the thickness of the metal being welded.

5. Make additional open butt joints, using plate spacings narrower and wider than the first. After welding, check the plates for bead appearance and penetration.

   Note: If the spacing between the plates is too great it may be necessary to run another bead over the first one to build the weld to desired dimensions. The first bead is referred to as the burning-in or root-pass bead, and the second bead is called the finish bead.

6. Once beads of good appearance can be made consistently, make additional joints, but weld from both sides. Check these welds by cutting the plates in two and examining the cross section for holes and *slag inclusions* (nonmetallic particles trapped in the weld).

REVIEW QUESTIONS

1. What can be done to prevent holes or slag inclusions in the weld?

2. What advantages does the open butt joint have over the closed butt joint?

3. Sketch a cross section of the weld made in step 2 showing penetration and fusion.

4. Sketch a cross section of the weld made in step 5, with the spacing too wide. Show what is wrong with this weld.

5. What is the space between two pieces of plate being welded called?

UNIT 48   SINGLE-PASS LAP JOINT

The welded lap joint has many applications and is economical to make since it requires very little preparation. For maximum strength it should be welded on both sides. A single pass or bead is enough for the plate used for this job. For heavier plate, several passes must be made.

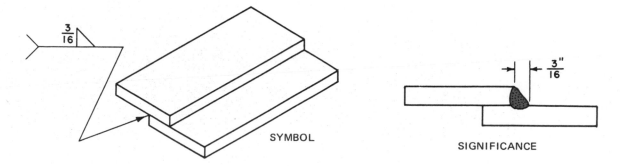

SYMBOL

SIGNIFICANCE

Fig. 48-1 Single-pass lap joint

Materials

Two steel plates, 3/16 inch thick, 2 in. x 9 in. to 12 in. each

1/8- or 5/32-inch diameter, E-6012 or E-6013 electrodes

Procedure

1. Set up the two plates as shown in figure 48-2. Make sure that the plates are reasonably clean and free of rust and oil. Be sure that the plates are flat and in close contact with each other.

2. Weld this joint with the electrode at an angle of 45 degrees from horizontal. Make sure that the weld metal penetrates the root of the joint, and that the weld metal builds up to the top of the lapping plate. Figure 48-3 shows the cross section, or end view, of a lap-welded joint, indicating the electrode angle and the size and

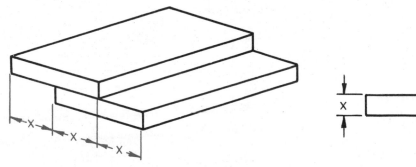

Fig. 48-2 Setup for lap joint

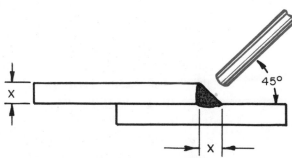

Fig. 48-3 Lap joint weld

Fig. 48-4 Properly welded lap joint

shape of the weld. Notice that the weld makes a triangle with each side equal to the thickness of the plate (See "X" in figure 48-3.)

3. Cool and clean this bead and examine it for uniformity. Pay particular attention to the line of fusion with the top and bottom plates. This should be a straight line with the weld blending into the plate, figure 48-4.

   Note: Too slow a rate of travel deposits too much metal and causes the weld to roll over onto the bottom plate. This forms a sudden change in the shape. The extra metal is a waste of material, and actually weakens the joint by causing stresses, figure 48-5.

4. Continue to make this type of joint until beads of uniform appearance can be made each time. Be sure to weld both sides of the assembly.

5. Make a test plate in the same manner as the other lap joints but weld it on only one side.

6. Place this plate in a vise so that the top plate can be bent or peeled from the bottom plate, figure 48-6. Bend this top plate until the joint breaks. Examine the break for penetration and uniformity. Another test may be made by sawing a lap-welded specimen in two and examining the cross section for penetration.

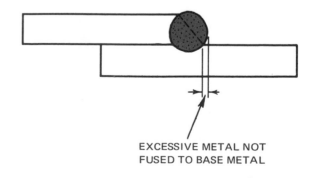

EXCESSIVE METAL NOT
FUSED TO BASE METAL

Fig. 48-5 Lap joint with excessive weld metal

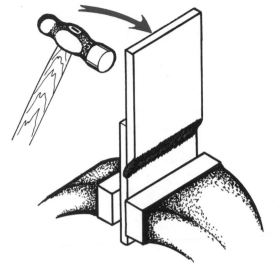

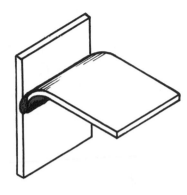

Fig. 48-6 Testing a lap weld

REVIEW QUESTIONS

1. Beginning students usually produce beads with an irregular line of fusion along the top edge of the overlapping plate. How is this corrected?

2. If the test shows lack of fusion at the root, how is this corrected?

3. What effect does too great a rod angle have on this type of joint?

4. What factor is most important in determining the location of the bead on a lap weld?

5. What is the result of depositing too much weld metal in a lap weld?

UNIT 49   SINGLE-PASS FILLET WELD

The fillet or T-weld is similar to the lap weld but the heat distribution is different. It has many industrial applications.

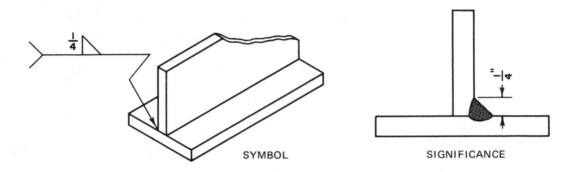

SYMBOL   SIGNIFICANCE

Fig. 49-1 Single-pass fillet weld

Materials

Two steel plates, 3/16 inch or 1/4 inch thick, 3 in. x 9 to 12 in. each

1/8- or 5/32-inch diameter, E-6012 electrodes

Procedure

1. Set up the two plates, figure 49-1. Be sure that the tack welds used to hold the plates in place are strong enough to resist cracking during welding, but not large enough to affect the appearance of the finished weld. This can be done by using a higher amperage and a higher rate of arc travel when tacking.

2. Make the fillet weld in much the same manner as the lap weld was made. The electrode angle is essentially the same. Both legs of the 45-degree triangle made by the weld must be equal to the thickness of the work for the full length of the joint.

3. Clean each completed weld and examine the surface for appearance. Look for poor fusion along both edges of the weld. Examine it for undercutting on the up-standing leg. If undercutting does exist, it is probably being caused by either too long an arc or too high a rate of travel.

4. When good fillet welds can be made every time, make a test weld on only one side of the joint. Then bend the top plate against the joint until it breaks, figure 49-2. Examine the break for root penetration and uniform fusion.

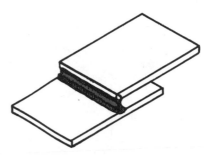

Fig. 49-2 Testing a fillet weld

REVIEW QUESTIONS

1. What can be done to prevent a hump at the start of the weld?

2. What causes undercutting?

3. What can be done to correct poor penetration and fusion in the root of the weld?

4. How is the size of a fillet weld measured?

5. If a fillet weld appears to be more on the flat plate than on the vertical plate how can it be corrected?

UNIT 50   MULTIPLE-PASS FILLET WELD

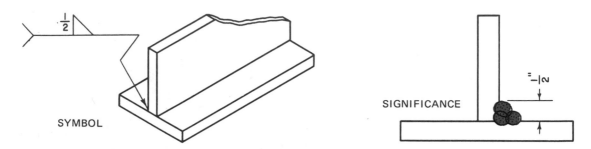

SYMBOL

SIGNIFICANCE

Fig. 50-1 Multiple-pass fillet weld

The heavy plate used in this unit requires 3 beads to complete the fillet.

Materials

Two steel plates, 1/2 inch thick, 2 in. x 9 to 12 in. each

5/32-inch diameter, E-6012 electrodes

Procedure

1. Set up and tack the plates in the manner used for the single-pass fillet welds.

2. Make a three-pass fillet weld, following the procedures shown in figures 50-1 and 50-2. Pay close attention to the angle of the electrode. When making the third pass, check the arc length frequently to make sure that an undercut does not develop along the upstanding leg of the weld.

3. Make additional fillet welds, welding both sides of the joint. Do not make all beads on one side of the joint first, but rather alternate the sequence.

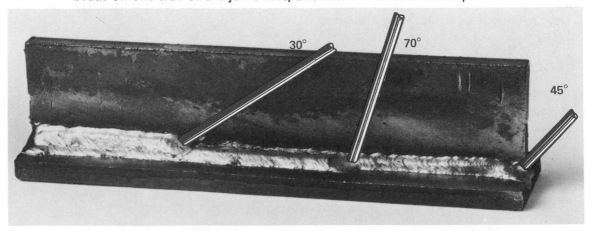

Fig. 50-2 Three-pass fillet weld using E-6012 electrode (right-handed operator).

4. Compare the distortion made by alternating sides with that made when only one side is welded.

5. Check the finished test plates in the same manner as the single-pass fillet welds were checked.

REVIEW QUESTIONS

1. What is the effect if the first bead on one side is followed by all three beads on the opposite side before completing the other two beads on the initial side? Why?

2. What should the height of the bead be for a fillet weld on 1/2-inch plate?

3. This unit indicates a weld made with three beads in two layers. How is a third layer applied?

4. How does the temperature of the base metal affect the overall appearance of the weld?

5. How does a hole in the first bead affect the second and third beads?

UNIT 51   WEAVING A LAP WELD

To lap-weld heavy plate with a single pass, a weaving motion of the rod is necessary. This requires a special technique to avoid an irregular, defective weld.

The weaving motion produces an oval pool of molten metal. The weld makes an angle of about 15 degrees with the edge of the plate.

The angle of the puddle varies with the amount of current, length of arc, and speed of welding as well as with the thickness of the welded metal. In general, larger welds at higher amperages need an angle slightly greater than 15 degrees.

● DENOTES PAUSE
IN ARC MOTION

Fig. 51-1 Weaving a lap weld in heavy plate

Materials

Two steel plates, 3/8 inch thick, 2 in. x 9 to 12 in. each

5/32-inch diameter, E-6012 electrodes

Procedure

1.   Position the plates for a lap weld.

2.   Weld the joint using a weaving motion to keep the bottom edge of the molten puddle ahead of the top of the puddle.

3.   When making this type of weld, pause with the arc at the top of each weave motion but not at the bottom of each weave, figure 51-1. This method prevents burning away or undercutting of the top of the weld.

4.   Cool, clean and examine the face of the finished weld. Look for undercutting of the top leg and overlapping of the bottom leg of the weld.

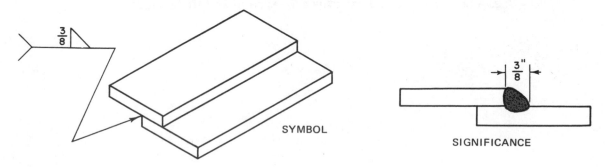

SYMBOL

SIGNIFICANCE

Fig. 51-2 Weaving a lap weld

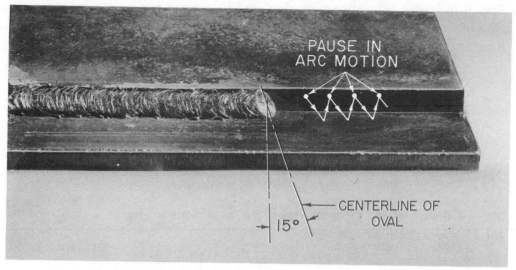

Fig. 51-3 Weaving a lap weld

5. Break this test plate and examine the root of the weld for good penetration.

6. Make another lap weld with the centerline of the molten puddle at a 90-degree angle to the line of the weld. Compare the line of fusion of the bottom leg of the weld with that of the previous bead.

7. Make more welds and examine them for appearance. All differences in ripple shape and spacing are caused by differences in arc control.

REVIEW QUESTIONS

1. Other than pausing at the top of each weave, how can irregularities and undercutting be controlled?

2. Does the undercutting referred to in this unit have the same appearance as the undercutting of a fillet weld?

3. Step 6 indicates an electrode motion with no lead at the bottom of the weave cycle. How does this affect the appearance of the finished bead and the line of fusion with the bottom plate?

4. Make a sketch of the cross section of the weld made in step 6. Show the correct shape for a lap weld with a dotted line.

5. How does the time for weaving a weld compare with that for a multiple-pass weld?

UNIT 52   WEAVING A FILLET WELD

This unit provides more practice in weaving beads to produce a multiple-pass weld of large size.

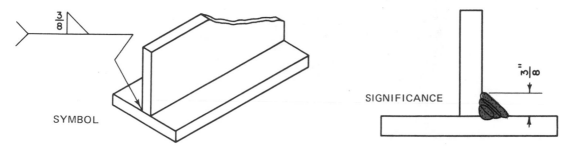

Fig. 52-1 Weaving a fillet weld

Materials

Two steel plates 3/8 inch thick, 2 in. x 9 to 12 in. each

5/32-inch diameter E-6012 electrodes

Procedure

1. Set up the plates as in figure 52-1.

2. Weld the joint using the electrode angle and weave motion described in unit 16. Weld a 3/8-inch fillet (i.e., each leg of the triangle formed by the weld should measure 3/8 inch).

   Note: The difference between a weave lap weld and a weave fillet weld is that the pause at the top of the puddle must be slightly longer for the fillet weld. Also, the length of the arc striking the up-standing leg must be kept very short to prevent undercutting.

3. Continue to make this type of weld. Examine each bead as it is made to determine what corrections are necessary to produce welds with uniform ripples and fusion.

4. Weave another bead over the original bead, using a very short arc and a definite pause as the arc is brought against the upstanding leg. Then bring the arc toward the bottom plate at a normal speed so that the bottom of the puddle leads the top of the puddle by 15 to 20 degrees.   Return the arc to the top of the fillet rapidly with a rotary motion, figure 52-2.

5. Clean and inspect this bead for undercutting along the top leg of the weld and for poor fusion along the bottom edge. Also check for uniformity of the bead ripples.

6. Weave a third bead over the first two so that the bottom of the bead leads the top slightly more than 20 degrees.   As the arc reaches the bottom of the fillet, try hooking the bead by moving the arc along the bottom before returning to the top of the weld.

135

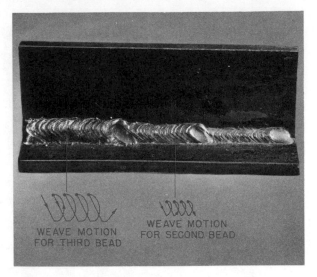

Fig. 52-2 Weave motions for a fillet weld

7. Clean and inspect weld as in step 5.

REVIEW QUESTIONS

1. What effect does hooking the bead have on the finished bead?

2. In making the second and third beads, is welding accomplished on the return stroke of the weave?

3. A common mistake in welding this type of fillet is depositing too much metal on the bottom plate. How is this fault corrected?

4. Is the bead made at step 6 good on 3/8-inch plate? Why?

5. Is undercutting more or less of a problem on the T joint than it is on the lap joint? Why?

UNIT 53   BEVELED BUTT WELD

This important joint can be very strong if it is well made. The multiple-pass procedure used in this unit is likely to produce a better joint than a thick single-pass method.

The beveled butt joint requires skill in weaving beads of two different widths, cleaning the preceding bead, and controlling the width of the bead.

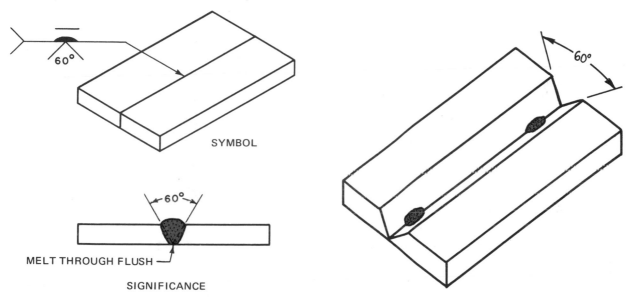

SYMBOL

MELT THROUGH FLUSH

SIGNIFICANCE

Fig. 53-1 Beveled butt weld   Fig. 53-2 Setup for beveled butt weld

Materials

Two steel plates, 3/8 inch thick, 4 in. x 9 to 12 in. each with one long edge beveled at 30 degrees

5/32-inch diameter E-6012 electrodes

Procedure

1. Align the plates on the welding bench and tack them as shown in figure 53-2.

2. Run a single-pass stringer bead in the root of the V formed by the two plates.

3. Clean the first bead and deposit a second bead by using a slight weaving motion. Allow the arc to sweep up the sides of the bevel in order to give this bead a slightly concave surface.

4. Clean the bead and run the third bead using a wide weaving motion. Do not allow the weld to become too wide. The actual width of the face of the weld should be slightly wider than the distance between the top edges of the V. Figure 53-3 is a cross-section view showing the size and contour of each bead.

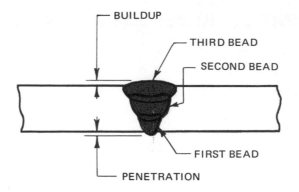

Fig. 53-3 Multiple-pass beveled butt weld

5. Clean and examine the finished weld for root penetration, evenness of fusion lines, and equal spacing of the ripples. Any variation in fusion, penetration, or ripple is caused by variations in arc manipulation. Uniform results can only be obtained by following uniform procedure.

    Note: Usually only three beads are required to make a bevel butt weld in 3/8-inch plate. However, if the first two beads are thin, do not attempt to make up for this by building the third bead much heavier. Instead, apply a normal third bead and a fourth if necessary. Heavy or thick buildups in one pass tend to produce holes and slag inclusions.

6. Align and tack a second set of plates as before, but leave an opening at the root of the V about one-half the rod diameter in size.

7. Weld these plates in the same manner as the first set and inspect visually.

REVIEW QUESTIONS

1. Why should the second bead have a slightly concave surface?

2. What effect does leaving a slight gap at the root of the V have on the finished bead?

3. If it is difficult to make the root pass with the plates gapped because of burn-through, what step is taken to correct this difficulty? Why?

4. How should the width of the finished bead compare with the width of the joint?

UNIT 54   OUTSIDE CORNER WELD

Outside corner welds are frequently used as finished corners after they have been smoothed by grinding or other means, figure 54-1. In this case, the shape of the bead and the smoothness of the ripples are very important. Roughness, caused by too much or not enough weld metal and uneven ripples, requires a lot of smoothing. This results in higher cost.

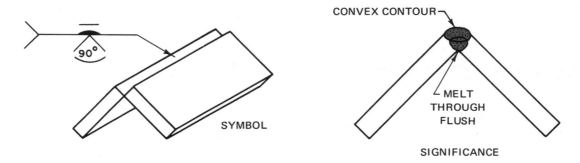

SYMBOL

CONVEX CONTOUR

MELT THROUGH FLUSH

SIGNIFICANCE

Fig. 54-1 Outside corner fillet

Materials

Two steel plates, 3/8 inch thick, 2 in. x 9 to 12 in. each

5/32-inch diameter E-6012 electrodes

Procedure

1. Set up the plates and tack them as shown in figure 54-2.

2. Make the weld in three or more passes as for a beveled butt joint. When making the final pass, observe all the precautions for weave welding to avoid any possibility of the finished bead overhanging the plate edges, figure 54-4. Place the assembly on the welding bench so that the weld can be made in the flat position.

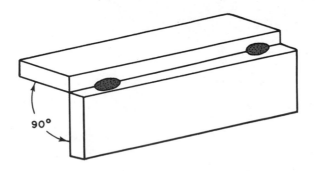

Fig. 54-2 Setup for outside corner weld

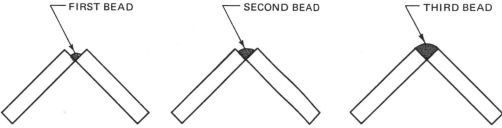

Fig. 54-3 Steps in multiple-pass corner weld

3. Check the finished weld for uniform appearance and to see if the angle of the plates is 90 degrees. Make any necessary corrections when setting up the next set of plates.

4. Test the weld by placing the assembly on an anvil and hammering it flat. Examine the root fusion and penetration, figure 54-5.

5. Make additional joints of this type, checking each weld for uniform appearance and for the shape shown in figure 54-4.

Fig. 54-4 Multiple-pass outside corner weld

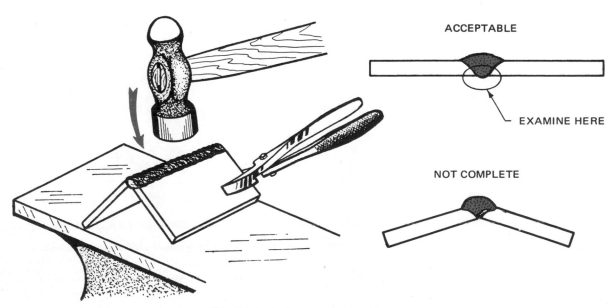

Fig. 54-5 Testing a corner weld

REVIEW QUESTIONS

1. How are poor fusion and uneven penetration corrected?

2. How are deep holes between the ripples in a finished weld corrected?

3. What can be done to prevent the final pass from building up too much, causing a hump along the line of fusion?

4. Is undercutting possible on an outside corner weld? How?

UNIT 55 OUTSIDE CORNER AND FILLET WELD
(HEAVY COATED ROD)

This unit provides an opportunity to gain skill and knowledge in the use of heavy-coated electrodes. These electrodes are recommended for use in the downhand or flat position only. This unit also provides experience in adjusting the electrode angle, arc length, and current setting.

Research and development have produced an electrode of this type with a large amount of iron powder added to the coating. This type of electrode makes a weld with good physical characteristics, and very good appearance and contour. The slag is so easily removed that it is described as self-cleaning.

The current values and arc voltages required to make this weld are quite high. As a result, the weld metal deposits rapidly. As much as 17 pounds of weld metal per hour may be deposited when using a heavy-coated 1/4-inch electrode.

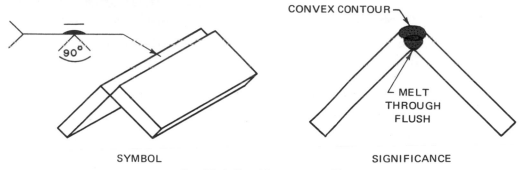

Fig. 55-1 Outside corner weld

Materials

Steel plate, 3/8 inch thick, 2 in. x 9 to 12 in.

5/32-inch diameter x 14-inch E-7024 electrode

Procedure

1. Set up and tack the plates as indicated.

2. Adjust the current according to the manufacturer's recommendations, or use the formula for current settings. For a heavy-coated rod, 20 percent should be added to the current value, so that the minimum setting is about 190 amperes.

3. Make a root pass with little or no weaving motion.

4. Clean and examine this bead for appearance and note the ease of slag removal. Also examine the end of the electrode and note that the metal core has melted back into the coating to form a deep cup.

   Note: The cupping characteristic makes this rod good for *contact welding.* In contact welding the coating is allowed to touch the work. In this manner the operator does not have to maintain the arc length. However, this limits the possibility of controlling

Fig. 55-2 Outside corner weld with heavy coated electrode

the rod, so the width of the weld is limited to stringer beads.
For this reason most experienced operators prefer to maintain a
free arc which can be controlled.

5. Apply a second bead with a weaving motion and a rate of travel that builds up a bead as thick as the plates being joined.

6. Clean and inspect the bead. Especially notice the line of fusion, appearance of ripples, and the shape of the finished joint. Figure 55-2 shows both beads.

7. Make a series of joints of this type, but increase the current for each joint by 10 to 15 amperes until a final joint is made at 250 amperes. Also vary the amount of *lead* (the angle at which the electrode points back toward the finished bead).

8. Clean and inspect the joints. Hammer each joint flat on an anvil. Examine the weld metal for holes and slag inclusions, and compare the grain sizes.

   Note: Valuable practice in making fillet welds with the E-7024 electrode is gained by using the plates welded into outside corner welds, steps 1 through 7.

9. Set up the assembly to weld the inside corner as shown in figure 55-3. Make sure that any slag, which may have penetrated this inside corner from the original welds, is thoroughly cleaned from the joint.

10. Adjust the current values as in step 2, and proceed to make a single-pass 3/8-inch fillet weld. The electrode should point back toward the finished weld at an angle of 15 to 20 degrees.

11. Clean and inspect this bead for the shape of the weld, straight, even fusion line between the weld and plate, and any evidence of holes or slag inclusions in the

Fig. 55-3 Fillet weld with heavy coated electrode

finished bead. Slag inclusions are usually caused by either poor cleaning of the original assembly, or a rod angle that caused the slag to flow ahead of the arc and molten pool.

12. Run a second and third bead on top of the first, using a uniform, slightly weaving motion to produce enough width. Weld other joints with current values up to 250 amperes.

13. Clean and inspect, as in step 11.

14. Set up plates and make a three-pass fillet weld as in steps 10 through 12, figure 55-3.

15. Break this weld by hammering the two plates together. Check for penetration and grain structure.

16. Weld more joints with one leg of the final assembly flat and the other vertical. Try varying the electrode angle from more than 45 degrees to less than 45 degrees from horizontal, and check the results for fusion and bead shape.

17. Make additional joints as in steps 10 through 12, but incline the plates so that the welding proceeds in a slightly downhill direction. Then make some joints with the weld proceeding in a slightly uphill direction.

18. Clean and inspect these joints and compare the results in each case with the welds made when the joint was in a perfectly flat position.

REVIEW QUESTIONS

1. What polarity does the manufacturer recommend for this rod?

2. It is suggested that various rod angles be tried while making these joints. What does this experiment show?

3. Too much current applied to a standard-coated rod causes the coating to break down and burn some distance up the rod. Is this true with the iron-powder type of electrode?

4. How do the higher amperages used in step 7 affect the surface appearance of the joint?

5. How does the speed of making these fillet welds compare with the speed of similar welds made with an E-6012 electrode?

6. What can be done to correct the tendency of the slag to run ahead of the arc and make holes in the bead?

7. How does the cleaning time for welds made in this unit compare with that for similar welds made with E-6010 and E-6012 electrodes?

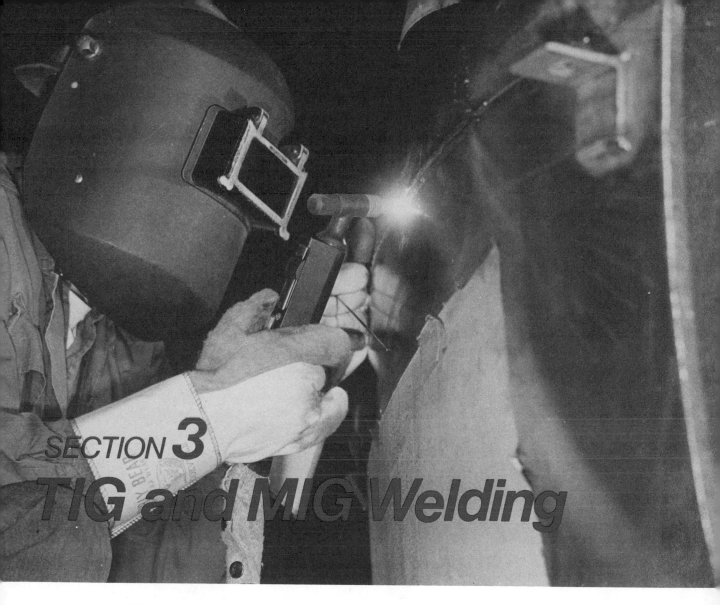

SECTION 3
TIG and MIG Welding

TIG (tungsten inert gas) welding is basically a form of arc welding in a controlled atmosphere. A non-consumable tungsten electrode creates an arc which passes through an inert atmosphere to produce a weldment with no oxidation or slag. It is especially useful in the welding of aluminum. Developed in the period of 1940 to 1960, it has rapidly become one of the indispensable welding methods.

The equipment used is more complex and expensive because electricity, water and the gas environment must all be provided and controlled. However, these additional expenses are offset by the utility of the TIG welding process.

In the MIG (metallic inert-gas) process, a consumable electrode in the form of wire is fed from a spool through the torch, often referred to as a welding gun. As the wire passes through the contact tube in the gun, it picks up the welding current.

MIG welding differs from TIG in that it is a one-handed operation. Therefore, it does not require the same degree of skill as the two-handed process, but it does maintain the utility of inert-gas shielding.

UNIT 56   THE TUNGSTEN INERT-GAS WELDING PROCESS

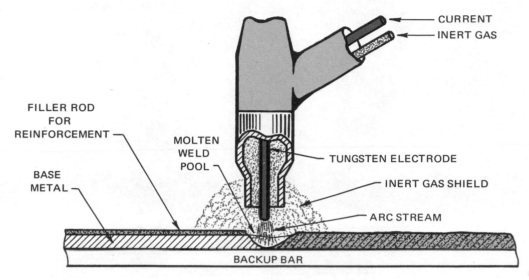

Fig. 56-1 TIG Welding process

The tungsten inert-gas shielded-arc welding process, figure 56-1, is an extension, refinement, and improvement of the basic electric arc welding process.

In the complete name of this process:

Tungsten refers to the electrode which conducts electric current to the arc.

Inert refers to a gas which will not combine chemically with other elements.

Gas refers to the material which blankets the molten puddle and arc.

Shielded describes the action of the gas in excluding the air from the area surrounding the weld.

Arc indicates that the welding is done by an electric arc rather than by the combustion of a gas.

The process is commonly referred to as *TIG welding* which is obtained from the first letter of each of the words, tungsten, inert, and gas. This type of welding is often referred to as Heliarc®, which is the trade name of a particular manufacturer. The TIG process generally produces welds which are far superior to those made by metallic arc welding electrodes.

ELEMENTS OF THE PROCESS

As shown in figure 56-2, the basic process uses an intense arc drawn between the work and a tungsten electrode. The arc, the electrode, and the weld zone are surrounded by an inert gas which displaces the air to eliminate the possibility of contamination of the weld by oxygen and nitrogen in the atmosphere. The tungsten electrode has a very high melting

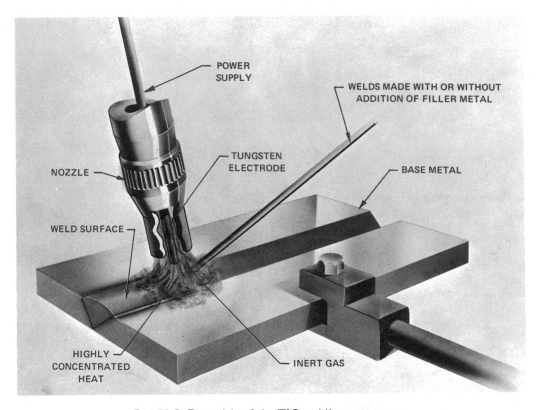

Fig. 56-2 Essentials of the TIG welding process

point (6,900 degrees F.) and is almost totally nonconsumable when used within the limits of its current-carrying capacity.

The inert gas supplied to the weld zone is usually either helium or argon, neither of which will combine with other elements to form chemical compounds. Argon gas is usually recommended because it is more generally available and better suited for use in the welding of a wide variety of metals and alloys. The basic components for a water-cooled TIG welding outfit are indicated in figure 56-3.

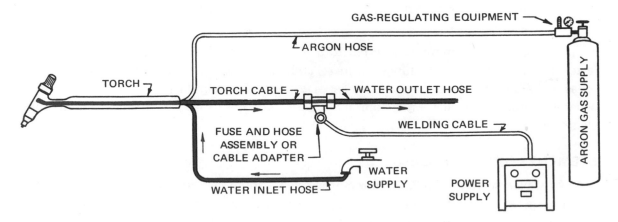

Fig. 56-3 Essentials for water-cooled TIG welding

ADVANTAGES OF TIG WELDING

Examples of the beads welded by arc, oxyacetylene and TIG processes are shown in figure 56-4.

- No flux is required and finished welds do not have to be cleaned of corrosive residue. The flow of inert gas keeps air away from the molten metal and prevents contamination by oxygen and nitrogen.

- In the chemical composition, the weld itself is usually equal to the base metal being welded. It is usually stronger, more resistant to corrosion, and more *ductile* (ability of a metal to deform without fracturing) than welds made by other processes. The inert gas will not combine with other elements or permit contamination by such elements, thus keeping the metal pure.

- Welding can be easily done in all positions. There is no *slag* (waste material entrapped in weld) to be worked out of the weld.

- The welding process can be easily observed. No smoke or fumes are present to block vision, and the welding puddle is clean.

- There is minimum distortion of the metal near the weld. The heat is concentrated in a small area and thus tends to minimize stresses.

- There is no splatter to cause metal-cleaning problems. Since no metal is transferred across the arc, this problem is avoided.

ARC WELDING

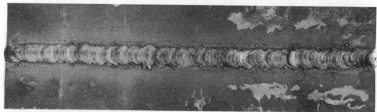

OXYACETYLENE WELDING

TIG WELDING

Fig. 56-4 Comparison of beads as welded

- Practically all the metals and alloys used industrially can be fusion-welded by the TIG process in a wide variety of thicknesses and types of joints.

- TIG welding is used particularly for aluminum and its alloys (even in very thick sections), magnesium and its weldable alloys, stainless steel, nickel and nickel-base alloys, copper and copper alloys, some brasses, low alloy and plain carbon steel, and the application of hard-facing alloys to steel.

REVIEW QUESTIONS

1. What does the term TIG welding refer to?

2. What are the essentials of TIG welding?

3. How does a TIG weld compare chemically with a metallic arc weld?

4. How do the mechanical properties of TIG welds compare with those of welds made by other manual processes and with the base metal?

5. What is meant by the term slag entrapment? Why is it harmful?

6. How do the uses of TIG welding compare with other manual processes?

7. Why is argon recommended as the shielding gas for most TIG welding?

UNIT 57 EQUIPMENT FOR MANUAL TIG WELDING

The equipment and material required for TIG welding consist of an electrode holder, or torch, containing gas passages and a nozzle for directing the shielding gas around the arc; nonconsumable tungsten electrodes; a supply of shielding gas; a pressure-reducing regulator and flowmeter; an electric power unit; and on some machines a supply of cooling water.

THE TORCH

A specially designed torch is used for TIG welding. It is so constructed that various sizes of tungsten electrodes can be easily interchanged and adjusted. The torch is equipped with a series of interchangeable gas cups to direct the flow of the shielding gas. Some of the torches are air-cooled, but water-cooled torches are more widely used.

SOURCE OF ELECTRIC CURRENT

The source of the electric current used in modern TIG welding is a specially designed welding machine, figures 57-1 and 57-2. It is possible to adapt standard alternating-current (AC) and direct-current (DC) welding machines such as are used in arc welding operations to TIG welding operations. However, a unit of this type is bulky and hard to manage when compared with the modern machines that are designed for TIG welding. An AC arc is best

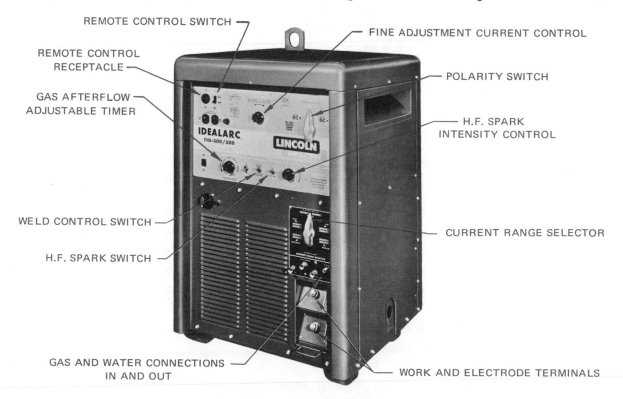

Fig. 57-1 AC - DC TIG welding machine

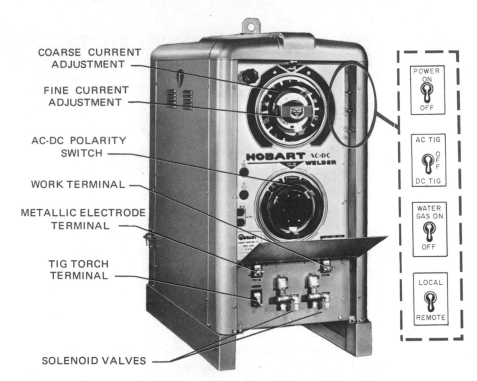

COARSE CURRENT ADJUSTMENT

FINE CURRENT ADJUSTMENT

AC-DC POLARITY SWITCH

WORK TERMINAL

METALLIC ELECTRODE TERMINAL

TIG TORCH TERMINAL

SOLENOID VALVES

Fig. 57-2 AC - DC TIG welding machine

suited for aluminum and some other metals and alloys. But a standard 60-cycle alternating current, which changes its direction of flow 120 times a second, is unsuited for welding because the electrical characteristics of the oxides on these metals cause the arc to extinguish (go out) at every half cycle or change of direction.

If, however, an *igniter arc current* is added to the standard 60-cycle current, the tendency to extinguish will be overcome because the igniter current will maintain a path for the standard 60-cycle current to follow. This igniter current is usually generated within the machine by a spark-gap oscillator which causes the current to change direction, not 120 times a second, but millions of times each second. Because this frequency of change is very high, the term high frequency is used in describing this current. Since standard 60-cycle alternating current actually does the welding, it is called AC welding; hence the term, *high-frequency alternating current* welding or, as it is commonly referred to, HFAC.

TIG welding power units, in most cases, can supply either AC or DC power to the electrode. These welding machines are equipped with a high-frequency oscillator which injects the high-frequency igniter current into the welding circuit. This high-frequency current causes a spark to jump from the electrode to the work without contact between the two. This ionizes the gap and allows the welding current to flow across the arc. Some manufacturers provide an external means of varying the frequency of this alternating current. In other machines the housing must be opened to make adjustments.

CONTROLS

Figure 57-3 shows the control panel for an AC-DC welding machine. TIG power units are usually equipped with solenoid valves to turn the flow of shielding gas and cooling water

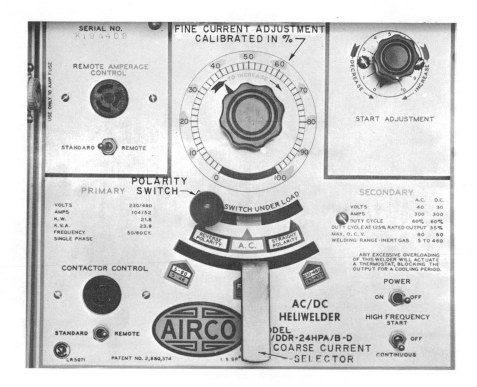

Fig. 57-3 Control panel for AC - DC welding machine

on and off. They are also provided with a remote-control switch, either hand- or foot-operated, to turn the water and gas on and off. Some of these remote-control devices also turn the main welding current on or off at the same time.

Most manufacturers equip the solenoid valves with a delayed-action device which allows the cooling water and shielding gas to continue to flow after the remote-control switch has been set at the stop or open position. This delay allows the tungsten electrode to cool to the point that it will not oxidize when the air comes in contact with it.

Some machines have an external means of varying the time of this afterflow to correspond to the electrode which is being used. In other types of machines, the housing must be opened to make this adjustment, figure 57-4. In any case, the shielding gas must be allowed to flow long enough so that the tungsten electrode cools until it has a bright, shiny surface.

SHIELDING GAS

The shielding gases are distributed in standard cylinders, which contain 330 cubic feet at 3,000 p.s.i.

As with all compressed gases, a regulator must be provided to reduce the high cylinder pressure to a safe, usable working pressure.

The main difference between the regulators used for oxyacetylene welding and those used for TIG welding is that the working pressure on the oxyacetylene regulators is indicated in pounds per square inch while the regulators used for TIG welding indicate the flow of shielding gas in cubic feet per hour. The latter are generally referred to as *flowmeters*. A combination regulator and flowmeter is shown in figure 57-5.

Fig. 57-4 Lower portion of machine: cover removed

Fig. 57-5 Combination regulator and flowmeter

Another significant difference between a standard regulator and the flowmeter is that the regulator will indicate the working pressure to the torch regardless of the regulator's position, while the tube on the flowmeter must be in a vertical position if an accurate reading is to be obtained.

REVIEW QUESTIONS

1. How long should the shielding gas and cooling water be allowed to flow after the welding arc is broken?

2. How do modern TIG welding machines compare with earlier models?

3. What makes the modern TIG torch adaptable to a wide range of welding operations?

4. What precaution is used to install a flowmeter on a gas cylinder?

5. In the air-cooled TIG torch, what does the cooling?

UNIT 58   THE WATER-COOLED TIG WELDING TORCH

The TIG torch, figure 58-1, is a multipurpose tool. It serves as:

- A handle.
- An electrode holder.
- A means of conveying shielding gas to the arc.
- A conductor of electricity to the arc.
- A method of carrying cooling water to the torch head.

COOLING WATER FLOW

Figure 58-2 shows a cross section of a torch with the cooling water flow indicated. The water cools the torch head, the collet and the electrode; it also cools the relatively light welding current cable which will overheat and burn if it is not surrounded by cooling water at

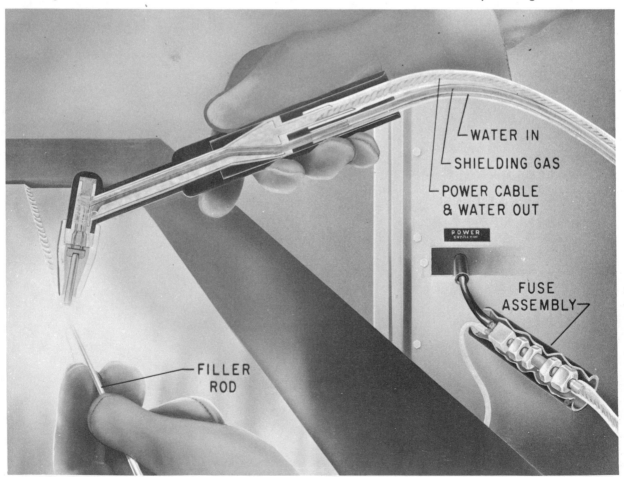

Fig. 58-1 Water-cooled TIG welding torch

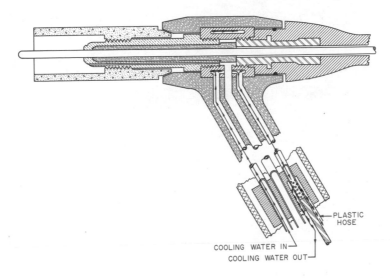

Fig. 58-2 Cooling water circuit

all times when current is flowing. All equipment manufacturers will supply cooling water requirements. A typical recommendation for cooling a medium-duty (300 amp.) torch is one quart of water per minute at 75 degrees F. (24 degrees C) or less, at not over 50 pounds per square inch pressure.

THE FLOW OF GAS

Figure 58-3 indicates the path of the regulated shielding gas through a hose to the torch head and through the collet holder. The gas then flows through a series of holes around the collet holder which direct the flow around the tungsten electrode through the ceramic nozzle to the work zone. The diameter and length of this nozzle vary with the size of the electrode used, the type of current being used, the material being welded, and the shielding gas being used.

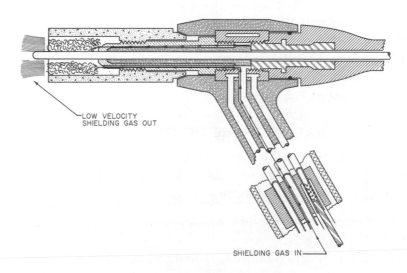

Fig. 58-3 Shielding gas flow

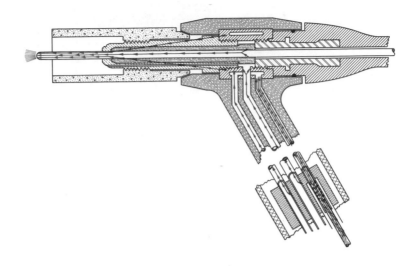

Fig. 58-4 Electric circuit

THE FLOW OF ELECTRICITY

As shown in figure 58-4, the electric current flows through the water-cooled welding cable, through the torch head to the collet holder and collet, to the tungsten electrode which forms one terminal of the arc, then through the work, and back through the ground cable to the power source. The student should examine a torch and compare the size of this ground cable with that of the cable leading into the torch.

The description of the electric circuit through the torch and work indicates that the electrode is negative or on straight polarity. If reverse polarity is used, the current flow is in the opposite direction.

If 60-cycle alternating current is used, the direction of flow changes 120 times each second so the electrode is positive (+) 60 times each second and negative (–) 60 times each second. This alternating of welding current is found to be very advantageous in many TIG welding operations. Direct-current straight polarity is referred to as DCSP; direct-current reversed polarity, as DCRP; and alternating current as AC, or in the case of AC with a superimposed high-frequency igniter current, as HFAC. The student should learn and remember these abbreviations since they are used in nearly every technical manual and paper written on the subject of TIG welding.

THE ASSEMBLY AND OPERATION OF THE TIG TORCH

To prepare the torch, figure 58-5, for welding operations, first choose the proper size electrode, matching collet and gas cup or nozzle. Some torches also require that the collet holder be changed with each different collet. Figure 58-6 shows a TIG torch with inter-, changeable collets.

1. Remove the collet cap or gas cap.

2. Remove the nozzle or gas cup by turning it counterclockwise.

3. If the collet holder is removable, take it off by turning it in a counterclockwise direction.

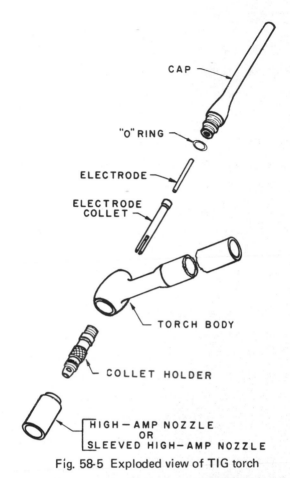

CAP

"O" RING

ELECTRODE

ELECTRODE
COLLET

TORCH BODY

COLLET HOLDER

HIGH—AMP NOZZLE
OR
SLEEVED HIGH—AMP NOZZLE

Fig. 58-5 Exploded view of TIG torch

COLLET CAP ASSEMBLY
FOR 2" ELECTRODES

COLLET CAP ASSEMBLY
FOR 7" ELECTRODES

COLLETS

COLLET CAP ASSEMBLY
FOR 3" ELECTRODES

ALUMINA SHORT
NOZZLE

ALUMINA STANDARD
NOZZLE

HOLDER BODY

Fig. 58-6 TIG torch with interchangeable collets

Note: In the case of a torch with a single collet holder, step 3 can be ignored and the collet and electrode can be removed from the gas cup side of the torch.

4. Remove the collet and the electrode from the collet holder.

5. Choose the proper size electrode, collet holder, collet and gas cup or nozzle.

6. Screw the collet holder firmly into the torch.

7. Screw the nozzle into the collet holder firmly against the O ring on the torch body.

8. Place the proper collet in the collet holder and replace the gas cap or collet cap in the torch, leaving it loose by one-half to one turn.

9. Insert the electrode through the gas nozzle into the collet.

Note: Never insert the electrode in the collet before inserting the collet in the torch. This guards against inserting a used electrode in the collet and, after use, finding that each end of the electrode has a ball formed on the end thus preventing the removal of the electrode from the collet.

10. Adjust the electrode for the recommended extension beyond the nozzle and tighten the gas cap or collet cap until the electrode is firmly fixed in the torch.

11. When the electrode extension needs to be adjusted to compensate for the slow burn-off, loosen the gas cap and adjust the electrode, and then firmly tighten the cap. Check the electrode to be sure it is firmly seated in the collet.

CAUTIONS:

- The ceramic nozzles are brittle, expensive, and easily broken. Always handle them with great care.

- Any electrical connection that is not thoroughly tight will generate extensive heat and may ruin the torch. Be sure all collet holders, collets and electrodes are tight to avoid costly damage.

- If the nozzle or gas cap is loose, it is possible for the shielding gas to draw air into the torch and contaminate the electrode as well as the weld. Always make sure that these parts are tight and that all O rings are in place.

REVIEW QUESTIONS

1. How does the size of the water-cooled cable to the torch compare with the ground or work cable?

2. What result would be expected if the water-cooled cable were not supplied with the cooling water at all times while the welding operation is going on?

3. What effect will a loose electrode or collet have on the torch?

4. What effect does a loose gas cap or ceramic nozzle have on the electrode and work zone?

5. If the electrode collet is put in the collet holder upside down, what happens?

UNIT 59 FUNDAMENTALS OF TIG WELDING

Although it can produce outstanding results, the TIG welding process may be unnecessarily expensive. A careless operator can cause a major expense by damaging the equipment. In particular, the tungsten electrode and the ceramic nozzle are subject to misadjustment. Care must be taken with these parts, therefore, since they are important to the production of high-quality work at a moderate cost.

COMPARISON OF METALLIC ARC WELDING WITH TIG WELDING

The variables found in TIG welding are almost identical to those in metallic arc welding. These variables are

- Length of arc
- Rate of arc travel
- Amount of current
- Angle of electrodes

The student who is skilled in the metallic arc welding process is already familiar with these variables. This prior knowledge is helpful in the study of TIG welding.

The following differences in costs must be considered:

- The overall value of the equipment used in TIG welding is much higher. This includes the welding machine, the cable and hoses, as well as the torch, regulator and nozzles.

- The shielding gas used is much more expensive than gases used in most other welding processes. For example, helium and argon gases are far more expensive than acetylene.

- The electrodes used in TIG welding are much more expensive. Actually, these electrodes are consumed so slowly that the cost of electrodes per foot of weld is very slight. However, the student should realize that any waste of electrodes due to bending, breaking or the use of excessive current is a very expensive error.

- The material being welded is generally much more expensive and sometimes hard to obtain, especially in the sizes and alloys desired.

In general, the student of TIG welding should be aware of the costs involved and should follow an intelligent, rigid procedure to protect the equipment from costly damage and to avoid the waste of expensive materials.

ELECTRODES

Tungsten and tungsten alloys are supplied in diameters of .010 inch, .020 inch, .040 inch, 1/16 inch, 3/32 inch, 1/8 inch, 5/32 inch, 3/16 inch, and 1/4 inch. They are manufactured in lengths of 3 inches, 6 inches, 7 inches, 18 inches, and, in some instances, 24 inches. The electrodes are made with a cleaned surface, either chemically cleaned and etched, or with a ground finish which holds the diameter to a closer tolerance. Electrodes are supplied in pure tungsten and in three alloys: 1 percent thorium, 2 percent thorium alloy and zirconium alloy.

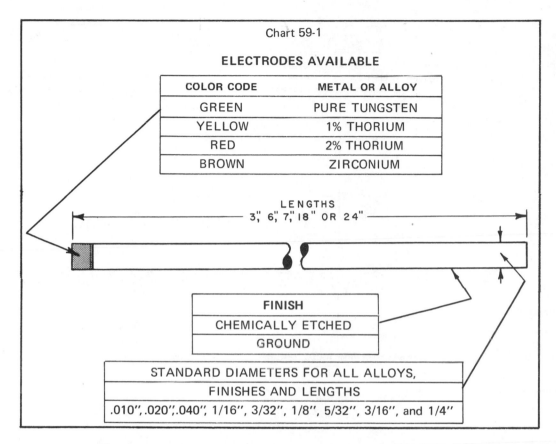

Chart 59-1

ELECTRODES AVAILABLE

COLOR CODE	METAL OR ALLOY
GREEN	PURE TUNGSTEN
YELLOW	1% THORIUM
RED	2% THORIUM
BROWN	ZIRCONIUM

LENGTHS
3", 6", 7", 18" OR 24"

FINISH
CHEMICALLY ETCHED
GROUND

STANDARD DIAMETERS FOR ALL ALLOYS,
FINISHES AND LENGTHS
.010", .020", .040", 1/16", 3/32", 1/8", 5/32", 3/16", and 1/4"

Pure tungsten is generally used with AC welding. The thoriated types are mostly used for DCSP welding and give slightly better penetration and arc starting characteristics over a wider range of current values. The zirconium alloy is excellent for AC welding and has high resistance to contamination. Its chief advantage is that it can be used in those instances when contamination of the weld by even very small quantities of the electrode is absolutely intolerable.

Chart 59-1 condenses information on electrodes available for TIG welding. The color code shown is being used by the major producers and distributors of tungsten electrodes.

The recommended amperage for any given size electrode varies with the type of joint being welded and the type of current used. A general recommendation when welding with AC is that the current be equal to the diameter of the electrode in thousandths of an inch multiplied by 1.25. For example, an electrode with a diameter of .040 requires a current of 40 x 1.25 or 50 amperes. Of course, the size of the electrode is a function of the thickness of the metal being welded. Chart 6-2 gives current ratings and electrode sizes for butt welding the various thicknesses of aluminum using HFAC arc with argon shielding and pure tungsten electrodes.

While charts are valuable as guides, a degree of sound judgment on the part of the operator is also desirable. Electrodes operated at a current value which is too low cause an erratic arc just as with metallic arc welding. If the current is correct, the end of the electrode appears as in figure 59-1.

Chart 59-2						
DATA FOR STRINGER BEADS IN ALUMINUM						
Thickness in Inches	HFAC Welding Current Flat* Amperes	Tungsten Electrode Diameter	Welding Speed Inches per Min.	Filler Rod Diameter	Recommended Argon Flow Cu. Ft. per Hr. ***	Gas Nozzle Size
1/16	60-80	1/16	12	1/16	15 to 20	4, 5, 6
1/8	125-145	3/32	12	3/32 or 1/8	17 to 25	6, 7
3/16	190-220	1/8	11	1/8	21 to 30	7, 8
1/4	260-300**	3/16	10	1/8 or 3/16	25 to 35	8, 10
3/8	330-380**	3/16, 1/4	5	3/16 or 1/4	29 to 40	10
1/2	400-450**	3/16, 1/4	3	3/16 or 1/4	31 to 40	10

* — Current values are for flat position only. Reduce the above figures by 10% — 20% for vertical and overhead welds.

** — For current values over 250 amps., use a torch with a water-cooled nozzle.

*** — Use lower argon flow for flat welds. Use higher argon flow for overhead welds.

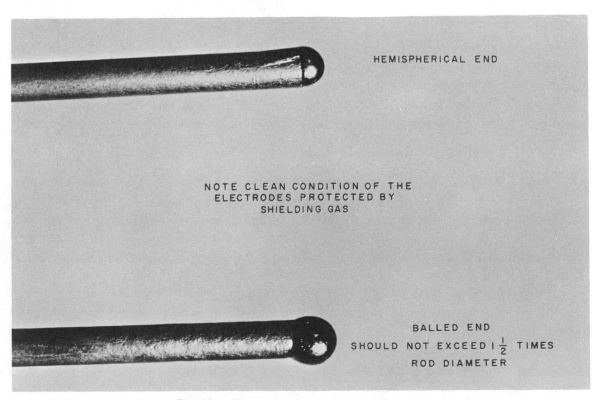

HEMISPHERICAL END

NOTE CLEAN CONDITION OF THE ELECTRODES PROTECTED BY SHIELDING GAS

BALLED END
SHOULD NOT EXCEED $1\frac{1}{2}$ TIMES
ROD DIAMETER

Fig. 59-1 Electrodes operated at proper current

Note that this figure shows a round, shiny end in one case and an end which forms a ball in the other. If this ball is over one and one-half times the diameter of the electrode, the current is too high and the electrode is consumed at an excessively high rate.

CERAMIC NOZZLES

The ceramic nozzles in chart 59-2 are indicated in fractions of an inch to describe the recommended inside diameter of these nozzles. There is a trend in the industry to indicate the nozzles by numbers such as 4, 5, 6, and 7. These numbers give the nozzle size in sixteenths of an inch. For example, a #6 nozzle indicates 6/16-inch or 3/8-inch inside diameter.

In general, the inside diameter or orifice of the nozzle should be from four to six times the diameter of the electrode. Nozzles with an orifice which is too small tend to overheat and either break or deteriorate rapidly. Smaller-diameter nozzles are also more subject to contamination. Ceramic nozzles are usually recommended for currents up to 250-275 amps. Above this point, special torches with water-cooled metal nozzles are generally used.

One other important factor is the amount the electrode extends beyond the nozzle, figure 59-2.

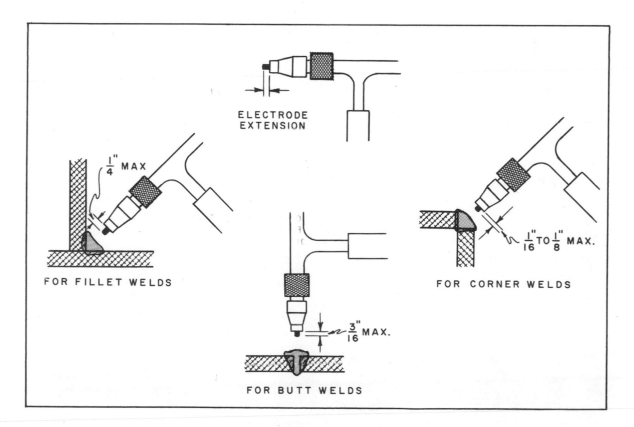

Fig. 59-2 Electrode extension

REVIEW QUESTIONS

1. Zirconium electrodes cost fifty percent more than tungsten but have excellent characteristics for HFAC welding. Would it be justifiable to use them?

2. Chart 59-2 shows the shielding gas flow for various sizes of electrodes. When welding aluminum would it be justifiable to experiment with this factor? On what basis?

3. Why would it be desirable to use seven-inch tungsten electrodes whenever possible instead of three-inch electrodes?

4. When would it be justifiable to use three-inch tungsten electrodes?

5. What is the objection to a long electrode extension?

6. What current in amperes is required for a 1/16-inch electrode?

UNIT 60   STARTING AN ARC AND RUNNING
STRINGER BEADS ON ALUMINUM

Rigid attention to detail and procedure is of extreme importance in TIG welding. Errors due to carelessness may prove to be very expensive. For instance, failure to turn on the cooling water usually results in destruction of the torch as well as the cable and hose assembly. Striking an arc with the machine set for normal amperage but with the polarity selector on DCRP will result in destruction of the electrode and usually the collet holder and collet.

Materials

Clean aluminum plate 1/8 in. thick x 4 in. x 6 in.
AC welding machine equipped with high-frequency oscillator
Cylinder of argon gas equipped with flowmeter
TIG torch fitted with 1/8-inch pure tungstem electrode and a #7 or #8 nozzle, (7/16 inch or 1/2 inch.)

Preweld Procedure

1. Make sure the torch is well away from the ground or work cable.

2. Turn on the cooling water.

3. Set the high-frequency switch to AC TIG.

4. Set water and gas switch to ON position.

5. If remote control is used, set switch to ON or remote. Otherwise leave it at LOCAL or OFF.

6. If the machine is equipped with a balanced wave filter or batteries, set this switch to ON.

7. Set gas and water afterflow timer for 1/8-inch electrodes.

8. Turn on the gas from the argon cylinder and adjust the flowmeter to supply 17 to 21 cubic feet per hour.

    Note: The flow of argon and the flow of water cannot be checked unless the remote control switch is ON or, if local control is used, the machine power switch is in the ON position.

9. Set the polarity switch to AC.

10. Adjust the current as indicated in chart 59-2.

11. Check the electrode for the proper extension. Refer to figure 59-2.

12. With the power OFF, check the electrode to be sure it is firmly held in the collet. To do this, place the exposed end of the electrode against a solid surface and push the torch down gently but firmly. If the electrode tends to move back into the nozzle, either the collet holder or gas cap needs to be tightened. Be sure to set the electrode extension and then tighten the collet holder or gas cap.

13. Turn the power switch ON. Turn the remote switch to ON. Note that the flowmeter indicates the proper flow of shielding gas. Check the waste line to be sure the cooling water is flowing.

14. Strike an arc by bringing the electrode close to a grounded workpiece, preferably copper. If the arc fails to jump from the electrode to the work without actual contact, set the high-frequency intensity control to a higher setting.

15. Again strike an arc on the copper and allow the current to flow until the electrode becomes incandescent. Break the arc and check the afterflow by watching the electrode cool. The instant the electrode becomes bright, look at the small ball in the flowmeter tube. It should drop to the bottom of the tube in two or three seconds. If it indicates a flow of a longer duration, adjust the afterflow timer to a slightly lower setting and repeat the above procedure until the timing is right.

   Note: If the afterflow timing was originally set for too short duration, the electrode would cool in the atmosphere and oxidize. This is indicated by the electrode becoming blue or black in color. In this event, adjust the afterflow timer higher until the ideal afterflow is reached.

The above procedure should be strictly followed each time the TIG welding process is used, even if someone else has just completed a weld with the equipment. It is the personal responsibility of each operator to be sure that all controls are properly adjusted at all times. Carelessness can lead to serious damage to the equipment.

Welding Procedure

1. Use all standard safety precautions. In addition, use a welding helmet filter plate one or two shades darker than the one for metallic arc welding.

2. Place the clean aluminum workpiece in firm contact with a clean worktable surface. The radio-frequency high-voltage (2000-4000 volts) igniter arc will jump a wide gap. If it jumps from the worktable to the reverse side of the workpiece, it may cause damage to the surface finish.

3. There are times when backing bars are advantageous. Either steel or stainless steel is recommended for backing bars; stainless steel is the better choice. Copper is not good for a backing bar when welding aluminum, and carbon should never be used for this purpose.

4. Turn on the remote switch and draw an arc with the electrode held as nearly vertical as possible while observing the molten puddle. Use an arc about equal in length to the electrode diameter.

5. Run a straight bead about 1/2 inch from, and parallel to, the edge of the plate, with the rate of the arc travel adjusted to maintain a pool of molten metal about 3/8 inch in diameter.

6. Examine the finished bead for uniformity and surface appearance. The weld should have a shiny appearance along its entire length. Note that there is an area about 1/8 inch on each side of the weld which is quite white but dull in lustre, figure 60-1. This

Fig. 60-1 Partially vaporized oxide, no preweld cleaning

is aluminum oxide which has been partially vaporized by the high-frequency igniter arc. Also examine the electrode for brightness and shape of the arc. Examine the reverse side of the plate for penetration.

7. Run a second bead 1/2 inch from the first and parallel to it. While making this bead, pay close attention to the area just outside the molten pool. A great number of pinpoint-size arcs should be seen. This is the high-frequency arc partially breaking up the aluminum-oxide film.

8. Examine the finished bead for surface appearance, uniformity, penetration and for the amount of white residue along the edges of the weld. Check the bead for cracking.

9. Make a third and fourth bead, observing the arc action and adjusting the rate of arc travel to obtain a uniform weld with complete penetration.

10. Examine these beads critically as in step 8. Examine the electrode after each bead for evidence of excessive burn-off or electrode contamination.

   Figure 60-2, view A, shows an electrode contaminated from the atmosphere because of too short a duration of afterflow of the shielding gas plus excessive electrode extension. In view B, an electrode is shown that was contaminated because of too short a duration of afterflow of shielding gas. View C shows an electrode contaminated by allowing it to come in contact with the molten pool.

11. Make a fifth bead but this time allow the electrode to contact the molten pool two or three times as the weld progresses. Observe the arc action and the area around the weld, especially after contact has been made.

12. Observe the finished bead as in step 7 and note the difference. Examine the electrode and gas nozzle for evidence of contamination as in figure 60-2, view C.

   Note: In the interest of electrode economy, the aluminum contamination of the electrode can be burned off by allowing the arc to dwell for several seconds on a copper plate. This is acceptable for practice welding. However, for high-production, high-quality welds, the accepted practice is to remove the electrode and either grind away the contaminated area or notch the electrode just back of the contamination and break it off. If this is done, be sure to grasp the electrode close to the notch to avoid bending.

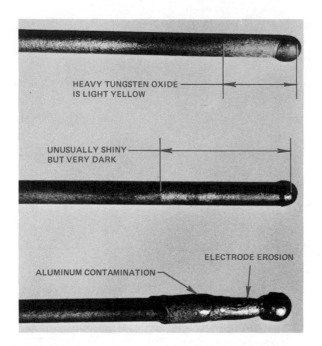

A. TOO LONG AN ELECTRODE
   EXTENSION PLUS TOO SHORT
   A DURATION OF AFTERFLOW

HEAVY TUNGSTEN OXIDE
IS LIGHT YELLOW

UNUSUALLY SHINY
BUT VERY DARK

B. SLIGHTLY TOO SHORT A DURATION
   OF AFTERFLOW

ALUMINUM CONTAMINATION

ELECTRODE EROSION

C. CONTAMINATION DUE TO CONTACT
   OF ELECTRODE AND MOLTEN
   PUDDLE

Fig. 60-2 Electrode Contamination

13. Do the experiment shown in figure 60-3. After the aluminum has broken cool the pieces and examine them. Observe that there is no evidence of melting. This indicates that the metal has broken rather than melted. This phenomenon, termed *hot-short,* is not uncommon. Many metals and alloys, such as copper, brass, and even cast iron display this characteristic at a temperature slightly below their melting point. Consider what would happen if the aluminum strip were laid on a flat steel plate and heated. Would the piece have broken or simply melted? What conclusion can be drawn from the above?

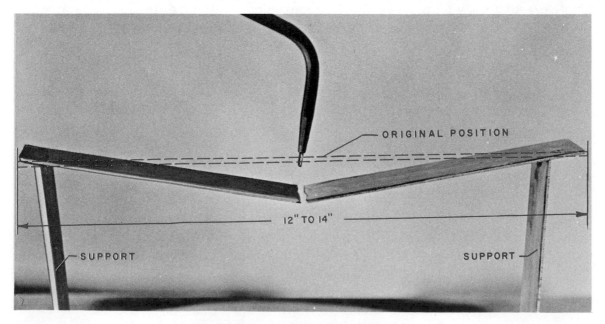

ORIGINAL POSITION

12" TO 14"

SUPPORT

SUPPORT

Fig. 60-3 Aluminum breaking under heat

REVIEW QUESTIONS

1. What four steps in the preweld procedure, in order of their importance, are most essential in avoiding damage to the equipment or electrodes?

2. For what two reasons would stainless steel be preferred over steel for a backing bar when welding aluminum?

3. What two reasons make copper a poor choice for a backing bar when welding aluminum?

4. Why would carbon be unacceptable for a backing bar when welding with tungsten electrodes?

5. Compare the amount of the white or light residue observed adjacent to the first weld with that observed in the succeeding welds. How do the welds differ?

6. What conclusion can be drawn from the observation in Question 5?

7. What practical use is there for the phenomenon observed in Questions 5 and 6?

8. What observations were made when welding with an electrode contaminated with aluminum?

9. What conclusions can be drawn from the answer to Question 8?

10. What conclusions can be drawn from the experiment in step 13?

UNIT 61   THE METALLIC INERT-GAS WELDING PROCESS

From an operator's viewpoint, it is easier to gain skill in the MIG process than in the TIG process. The deposition rate is much faster with MIG than TIG although the same metals can be joined with both. The thickness of material to be joined is a factor in choosing the correct process.

MIG welding (often called metal inert-gas or gas-metal arc welding) is done by using a consumable *wire electrode* to maintain the arc and to provide filler metal. The wire electrode is fed through the torch or gun at a preset controlled speed. At the same time, an *inert gas* is fed through the gun into the weld zone to prevent contamination from the surrounding atmosphere.

ADVANTAGES OF MIG WELDING

- Arc visible to operator
- High welding speed
- No slag to remove
- Sound welds
- Weld in all positions

TYPES OF MIG WELDING

- *Spray arc welding,* figure 61-1, is a high current range method which produces a rapid deposition of weld metal. It is effective in welding heavy-gage metals, producing deep weld penetration.

  At high currents, the arc stability improves and the arc becomes stiff. The transition point, when the current level causes the molten metal to spray, is governed by the wire type and size, and the type of inert gas used.

- *Short arc welding,* figure 61-2, is a reduced heat method with a pin arc for use on all common metals. It was developed for welding thin-gage metals to eliminate distortion, burn-through and spatter. This technique can be used in the welding of heavy thicknesses of metal.

- *MIG CO$_2$ (carbon-dioxide) welding* is a variation of the MIG process. Carbon dioxide is used as the shielding gas for

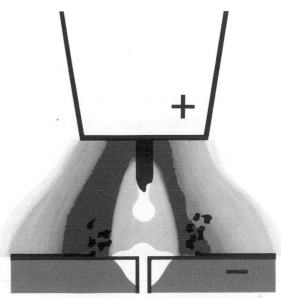

Fig. 61-1 Spray arc welding

Start of the short arc cycle — High temperature electric arc melts advancing wire electrode into a globule of liquid metal. Wire is fed mechanically through the torch. Arc heat is regulated by presetting the power supply.

Molten electrode moves toward workpiece. Note cleaning action. Argon gas mixture, developed specifically for short arc, shields molten wire and weld seam, insuring regular arc ignition, controlling spatter and weld contamination.

Electrode makes contact with workpiece, creating short circuit. Arc is extinguished. Metal transfer begins due to gravity and surface tension. Frequency of arc extinction in short arc varies from 20 to 200 times per second, according to "preset" conditions.

Molten metal bridge is broken by pinch force, the squeezing action common to all current carriers. Amount and suddenness of pinch is controlled by power supply. Electrical contact is broken, causing arc to reignite.

With arc renewed, short arc cycle begins again. Because of precise control of arc characteristics and relatively cool, uniform operation, short arc produces perfect welds on metals as thin as .030-in. with either manual or mechanized equipment.

Fig. 61-2 Short arc welding

the welding of carbon and low-alloy steel from 16 gage (.059 inch) to 1/4 inch or heavier. It produces deeper penetration than argon or argon mixtures with slightly more spatter. Carbon-dioxide MIG welding costs about the same as other processes on mild steel applications.

- *Cored-wire welding* is an intense-heat, high-deposition-rate process using flux-cored wire on carbon steel. Electrically, cored-wire welding is similar to spray-arc welding. In addition to inert-gas shielding, a flux contained inside the wire forms a slag that cleans the weld and protects it from contamination. In application, it is recommended for large fillet welds in the flat or horizontal position.

REVIEW QUESTIONS

1. What does the term MIG welding mean?

2. What is the principle of the MIG welding process?

3. What are four types of MIG welding?

4. What polarity is used for MIG welding?

5. What are the advantages of MIG welding?

UNIT 62 EQUIPMENT FOR MANUAL MIG WELDING

A specially designed welding machine is used for MIG welding. It is called a *constant voltage (CV) type* power source. It can be a DC rectifier or a motor- or engine-driven generator. (See figure 62-3.)

The output welding power of a CV machine has about the same voltage regardless of the welding current. The output voltage is regulated by a rheostat on the welding machine, figure 62-1. Current selection is determined by wire-feed speed. There is no current control as such.

The wire-feeding mechanism and the CV welding machine make up the heart of the MIG welding process, figures 62-2, 62-4, and 62-5. There is a fixed relationship between the rate of electrode wire burn-off and the amount of welding current. The electrode wire-feed speed rate determines the welding current.

The gun is used to carry the electrode wire, the welding current, and the shielding gas from the wire feeder to the arc area, figure 62-6. The operator directs the arc and controls the weld with the welding gun.

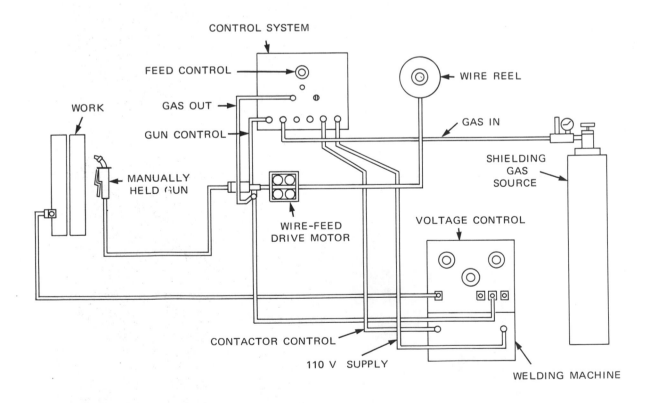

Fig. 62-1 MIG equipment

Fig. 62-2 Constant voltage rectifier

Fig. 62-3 Motor generator

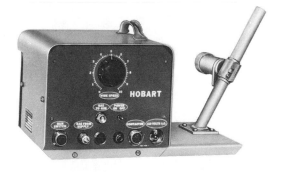

Fig. 62-4 Wire feed control unit

Fig. 62-5 Wire feed unit

SHIELDING GASES

The shielding gas can have a big effect upon the properties of a weld deposit. The welding is done in a controlled atmosphere.

Pure argon, argon-helium, argon-oxygen, argon-carbon dioxide, and carbon dioxide are commonly used with the MIG process. With each kind and thickness of metal, each gas and mixture affects the smoothness of operation, weld appearance, weld quality, and welding speed in a different way.

Gas-flow rate is very important. A pressure-reducing regulator and flowmeter are required on the gas cylinder. Flow rates vary, depending on types and thicknesses of the material and the design of the joint. At times two or more gas cylinders are connected (manifolded) together to maintain higher gas flow.

FILLER WIRES

The wire electrode varies in diameter from .030 inch to 1/8 inch. The composition of the electrode wire must be matched to the base metal being welded. In the welding of carbon steel, the wire is solid and bare except for a very thin coating on the surface to prevent rusting. It must contain deoxidizers which help to clean the weld metal and to produce sound, solid welds.

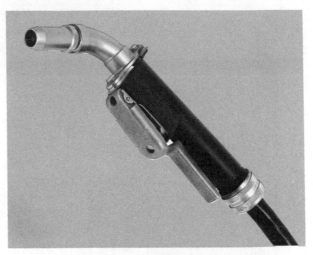

Fig. 62-6 Welding gun and cable assembly

REVIEW QUESTIONS

1. What are the main components of the MIG welding equipment?

2. What is considered to be the heart of the MIG welding process?

3. What does the wire-feed control determine?

4. What is the measurement of the flowmeter which registers gas flow to control the shielding atmosphere?

5. Is the voltage controlled by the wire feeder or the welding machine?

UNIT 63   MIG WELDING VARIABLES

Most of the welding done by all processes is on carbon steel. About 90 percent of all steel is plain carbon steel. This unit describes the welding variables in short-arc welding of 24-gage to 1/4-inch mild steel sheet or plate. The type of equipment usually found in training facilities lends itself well to these applications.

The applied techniques and end results in the MIG welding process are controlled by these variables and must be understood by the student. The variables are adjustments that are to be made to the equipment and also manipulations by the operator.

These variables can be divided into three areas.

- Preselected variables

- Primary adjustable variables

- Secondary adjustable variables

PRESELECTED VARIABLES

Preselected variables depend on the type of material being welded, the thickness of the material, the welding position, the deposition rate and the mechanical properties. These variables are

- Type of electrode wire
- Size of electrode wire
- Type of inert gas
- Inert-gas flow rate

Charts 63-1, 63-2, and 63-3 are references for the new MIG welding student. Manufacturers' recommendation also serve as a guide to be followed in these areas.

PRIMARY ADJUSTABLE VARIABLES

These control the process after preselected variables have been found. They control the penetration, bead width, bead height, arc stability, deposition rate and weld soundness. They are

- Arc voltage
- Welding current
- Travel speed

SECONDARY ADJUSTABLE VARIABLES

These variables cause changes in the primary adjustable variables which in turn cause the desired change in the bead formation. They are

- Stickout
- Nozzle angle
- Wire-feed speed

Chart 63-1

COMPARISON CHART
MILD STEEL ELECTRODES FOR MIG WELDING

MANUFACTURERS	American Welding Society Classification						A5-18-6		
	E 70S-1	E 70S-2	E 70S-3	E 70S-4	E 70S-5	E 70S-6	E 70S-G	E 70S-1B	E 70S GB
Airco Welding Products Div. Air Reduction Co. Inc.	S-20		A 675		A 666	A 681	A 608	A 608	A 608
Alloy Rods Company Div. Chemetron Corporation			MINIARC 70					MINIARC 100	
Hobart Brothers Company	TYPE 20		TYPE 25		TYPE 30	TYPE 28		TYPE 18	
Linde Div. Union Carbide Corporation	LINDE 29S	LINDE 65	LINDE 82, 66	LINDE 85		LINDE 86	LINDE 83	LINDE 83	
Midstates Steel & Wire Co.			IMPERIAL 75			IMPERIAL 88	IMPERIAL 95		
Modern Engineering Co. Inc.			MECO 60S-3		MECO 70S-5		MECO 70S-G		
Murex Welding Products			MUREX 1316		MUREX 1315		MUREX 1313 MO	MUREX 1313 MO	
National Cylinder Gas Div. Chemetron Corporation			MINIARC 70				MINIARC 100		
National Standard Company	NS-106	NS-103	NS-101			NS-115	NS-116	NS-102	
P & H Welding Products Unit of Chemetron Corporation			P & H CO-85		P & H CO-86		P & H CO-87		
Page, Division of ACCO	PAGE AS-20		PAGE AS-25		PAGE AS-30	PAGE AS-28		PAGE AS-18	

Chart 63-2

ALL JOINTS ALL POSITIONS MILD STEEL							
MATERIAL THICKNESS		NUMBER OF PASSES	WIRE DIAMETER	WELDING CONDITIONS DCRP		GAS FLOW CFH	TRAVEL SPEED IPM
GAGE	INCH			ARC VOLTS	AMPERES		
24	.023	1	.030	15-17	30-50	15-20	15-20
22	.029	1	.030	15-17	40-60	15-20	18-22
20	.035	1	.035	15-17	65-85	15-20	35-40
18	.047	1	.035	17-19	80-100	15-20	35-40
16	.059	1	.035	17-19	90-110	20-25	30-35
14	.074	1	.035	18-20	110-130	20-25	25-30
12	.104	1	.035	19-21	115-135	20-25	20-25
11	.119	1	.035	19-22	120-140	20-25	20-25
10	.134	1	.045	19-23	140-100	20-25	27-32
	3/16 in.	1	.045	19-23	180-200	20-25	18-22
	1/4 in.	1	.045	20-23	180-200	20-25	12-18

Chart 63-3

SHIELDING GASES FOR MIG		
METAL	SHIELDING GAS	APPLICATION
CARBON STEEL	75% ARGON 25% CO_2	1/8 inch or less thickness: High welding speeds without burn-through; minimum distortion and spatter
	75% ARGON 25% CO_2	1/8 inch or more thickness: Minimum spatter, good control in vertical and overhead position
	CO_2	Deeper penetration, faster welding speeds
STAINLESS STEEL	90% HELIUM 7.5% ARGON 2.5% CO_2	No effect on corrosion resistance, small heat-affected zone, no undercutting, minimum distortion
LOW ALLOY STEEL	60-70% HELIUM 25-35% ARGON 4-5% CO_2	Minimum reactivity, excellent toughness, excellent arc stability and bead contour, little spatter
	75% ARGON 25% CO_2	Fair toughness, excellent arc stability, and bead contour, little spatter
ALUMINUM, COPPER, MAGNESIUM, NICKEL AND THEIR ALLOYS	ARGON AND ARGON-HELIUM	Argon satisfactory on lighter material, Argon-helium preferred on thicker material

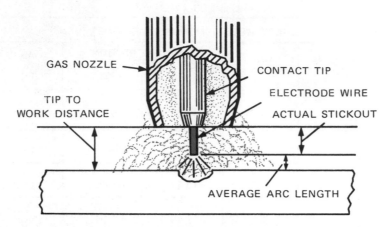

Fig. 63-1 Stickout

Stickout as shown in figure 63-1 is the distance between the end of the contact tip and the end of the electrode wire. From the operator's viewpoint, however, stickout is the distance between the end of the nozzle and the surface of the work.

Nozzle angle refers to the position of the welding gun in relation to the joint as shown in figure 63-2. The *transverse angle* is usually one-half of the included angle between plates forming the joints. The *longitudinal angle* is the angle between the centerline of the welding gun and a line perpendicular to the axis of the weld.

The longitudinal angle is generally called the nozzle angle and is shown in figure 63-3 as either trailing (pulling) or leading (pushing). Whether the operator is left-handed or right-handed has to be considered to realize the effects of each angle in relation to the direction of travel.

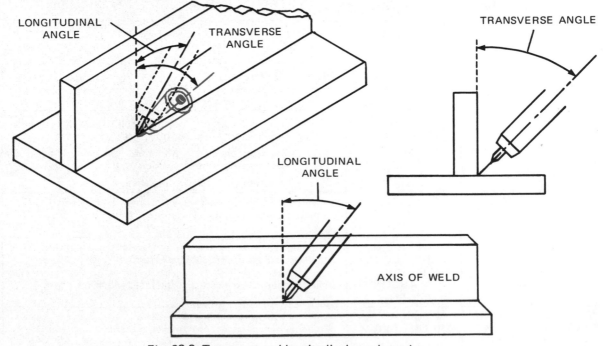

Fig. 63-2 Transverse and longitudinal nozzle angles

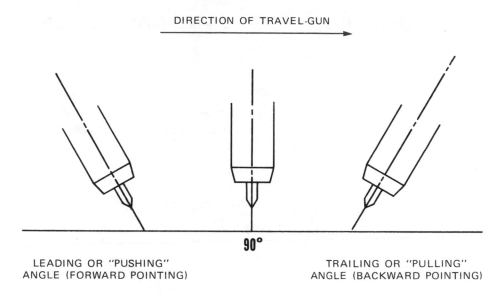

DIRECTION OF TRAVEL-GUN

90°

LEADING OR "PUSHING"
ANGLE (FORWARD POINTING)

TRAILING OR "PULLING"
ANGLE (BACKWARD POINTING)

Fig. 63-3 Nozzle angle, right-handed operator

REVIEW QUESTIONS

1. What must be considered before selecting the type and size of electrode wire and type of inert gas?

2. What controls the penetration and the bead width and height?

3. What is the distance between the end of the nozzle and the work called?

4. What are the two nozzle angles called?

5. Where does the electrode wire pick up its electrical current?

UNIT 64   ESTABLISHING THE ARC AND MAKING WELD BEADS

It is assumed that the welding equipment has been set up according to procedures outlined in the appropriate manufacturers' instruction manuals. Students should know how to perform adjustments and maintenance on MIG welding equipment. As in TIG welding, the equipment is expensive, and the student must realize that the equipment can be destroyed if instructions are not followed.

Materials

10-, 11-, or 12-gage mild steel plate 6 in. x 6 in.
.035-inch E 70S-3 electrode wire
CO_2 shielding gas

Preweld Procedure

1. Check the operation manuals for manufacturer's recommendations.

2. Set the voltage at about 19 volts.

3. Set the wire-feed speed control to produce a welding current of 110 to 135 amperes.

4. Adjust the gas-flow rate to 20 cubic feet per hour.

5. Recess the contact tip from the front edge of the nozzle 0 to 1/8 inch.

6. Review standard safe practice procedures in ventilation, eye and face protection, fire, compressed gas and preventive maintenance. Safety precautions should always be part of the preweld procedure.

Welding Procedure

1. Maintain the tip-to-work distance of 3/8 inch (stickout) at all times. See figure 63-1.

2. Maintain the trailing gun transverse angle at 90 degrees and the longitudinal angle at 30 degrees from perpendicular. See figures 63-2 and 63-3.

3. Hold the gun 3/8 inch from the work, lower the helmet by shaking the head, and squeeze the trigger to start the controls and establish the arc.

    Note: Operators should not form the habit of lowering the helmet by hand since one hand must hold the gun and the other may be needed to hold pieces to be tacked or positioned.

4. Make a single downhand stringer weld.

5. Practice welding beads. Start at one edge and weld across the plate to the opposite edge.

    Note: When the equipment is properly adjusted, a rapidly crackling or hissing sound of the arc is a good indicator of correct arc length.

6. Practice stopping in the middle of the plate, restarting into the existing crater and continuing the weld bead across the plate.

Note: When the gun trigger is released after welding, the electrode forms a ball on the end. To the new operator, this may present a problem in obtaining the penetration needed at the start. This can be corrected by cutting the ball off with wire cutting pliers. The ball can cause whiskers to be deposited at the start. *Whiskers* are short lengths of electrode which have not been consumed into the weld bead.

This procedure should be practiced often by the new operator. A satisfactory performance in welding joints depends on the ability to do this basic manipulation.

Checking Application

1. Examine the base metal to be sure it is free from oil, scale, and rust.
2. Recheck the equipment settings according to the operation manual. This includes gas flow, stickout, gun angle, arc voltage, and amperage from wire-feed speed.
3. Keep equipment clean, specifically the gun nozzle, feeder rolls, wire guides, and liners. An anti-spatter spray should be used on the nozzle to help keep it clean.

REVIEW QUESTIONS

1. What is a good indication of correct arc length when the equipment is adjusted properly?

2. Why is it necessary to have the stickout distance correct, especially at the start and end of the weld?

3. Why is spatter buildup on the inside of the nozzle harmful to a good weld?

4. What is a whisker in MIG welding?

5. Why is the ball end cut off the wire?

UNIT 65   MIG WELDING THE BASIC JOINTS

The ability to manipulate the equipment and apply the single bead across a piece of sheet or plate is the basis for the welding of various joints. Being able to see and follow a joint helps to insure equal fusion on both pieces.

> *Note:* Depending on the size of the gun, visibility may be a problem. The operator should be in the most comfortable position to see where the deposit is being made.

The welds that make up these basic joints can be applied in any position and can be single or multiple pass, depending on the thickness of the material being joined.

This unit describes the basics involved in welding the butt joint, lap joint, T joint, and corner joint in the flat position using 11-gage mild steel sheet. It is assumed that the operator can position and tack two pieces of this material to form these joints.

Materials

10-, 11-, or 12-gage mild steel sheet 1 1/2 in. x 6 in.
.035-inch E 70S-3 electrode wire
CO_2 shielding gas

Preweld Procedure

1. Check the operation manual for the manufacturers' recommendations.

2. Set the voltage at about 21 volts.

3. Set the wire-feed speed control to produce a welding current of 110 to 135 amperes.

4. Adjust the gas flow to 20 cubic feet per hour.

5. Adjust the voltage to get a smooth arc.

   > *Note:* Before attempting to weld a joint, always adjust the machine, using a piece of scrap material.

6. Recess the contact tip from the edge of the nozzle 0 to 1/8 inch.

7. Make sure that the equipment is clean; specifically the gun nozzle and liner, feeder rolls and wire guides.

8. Remember that safety should always be part of the preweld procedure.

Welding Procedure (Butt Weld)

1. Place two pieces on the worktable in good alignment and with two 6-inch edges spaced 1/16 inch apart (root opening), as in figure 65-1.

2. Maintain a stickout of 3/8 inch. See figure 63-1.

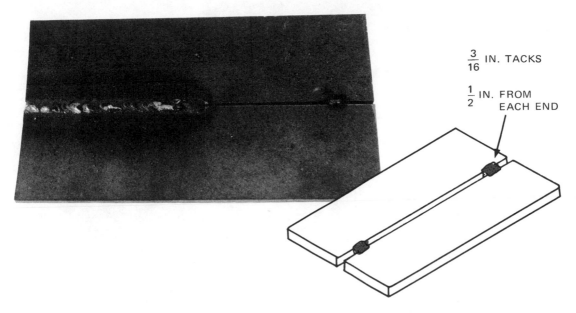

$\frac{3}{16}$ IN. TACKS

$\frac{1}{2}$ IN. FROM
EACH END

Fig. 65-1 Butt joint

3. Tack weld the two pieces together as shown in figure 65-1. The tacks should be placed about 1/2 inch from each end to avoid having too much metal and poor penetration at the start of the weld.

   Note: This method of tacking is used in all of the joints.

4. Use a transverse angle of 90 degrees or directly over and centered on the joint. Find the longitudinal angle by experimentation. The trailing gun angle is used first at 10 degrees perpendicular. The leading gun angle is used next at the same inclination. See figures 63-2 and 63-3.

   Note: Differences should be found and allowed for. The operator will prefer either the trailing gun angle or the leading gun angle, but the deciding factors are the degree of penetration, deposition rate, gas coverage of weld zone, and overall appearance.

5. Weld the joint on the tack side.

6. Cool the work and examine for uniformity.

7. Check the depth of penetration of the weld by first placing the assembly in a vise with the center of the weld slightly above and parallel to the jaws. Then bend the outstanding sheet toward the face of the weld. Penetration should be 100 percent with no faults.

8. Make more joints of this type and change the setting of the equipment slightly in different directions. Note what happens and what has to be done to compensate for it.

   Note: Thickness of material controls the root opening and whether or not the butt joint will need a bead on both sides.

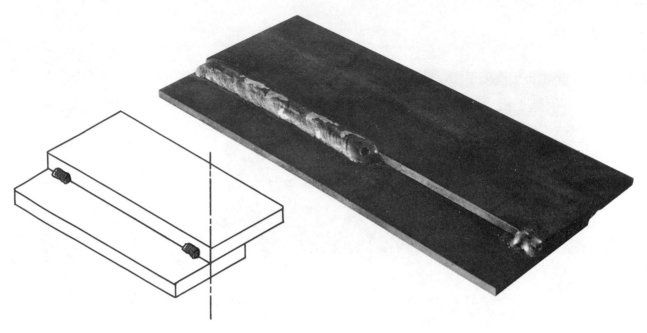

Fig. 65-2 Lap joint

Welding Procedure (Lap Joint)

1. Place two pieces on the worktable and tack as shown in figure 65-2.

2. Maintain a stickout of 3/8 inch. See figure 20-1.

3. The longitudinal angle is 10 degrees from perpendicular using a trailing gun angle. The transverse gun angle should be about 60 degrees from the lower sheet. The location of the nozzle in relation to the joint should be as shown in figure 65-3.

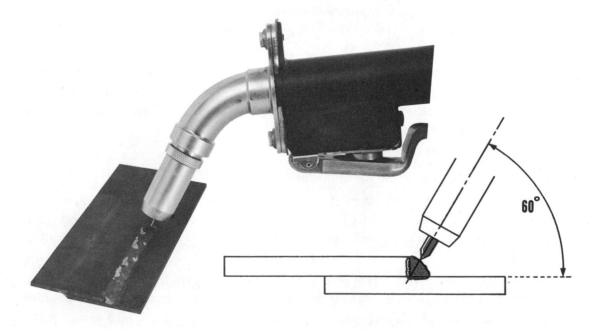

Fig. 65-3 Location of nozzle

Note: The operator's ability to compensate for the location of the gun in relation to the joint controls the uniformity of the bead and the desired amount of penetration.

4. Weld the joint. Allow for distortion by running the first bead on the opposite side from the tacks.

5. Cool and examine the bead for uniformity. Examine the line of fusion with the top and bottom sheets. This should be a straight line with no undercut.

6. Weld another lap joint on only one side (tack side). Place this piece in a vise in such a way that the top sheet can be bent from the bottom sheet 180 degrees, if possible, to check penetration and strength.

7. Make another test by sawing a lap-welded specimen in two and examining the cross section for penetration.

8. Make more joints of this type until they are uniform and consistent.

Welding Procedure (T Joint)

1. Place two pieces on the worktable and tack weld as shown in figure 65-4.

   Note: Hold one piece while tacking. This is good experience, as it is necessary to handle and manipulate the gun with one hand.

2. Maintain a stickout of 3/8 inch. See figure 63-1.

Fig. 65-4 T joint

3. The longitudinal angle is 10 degrees from perpendicular using a trailing gun angle. The transverse gun angle should be about 45 degrees from the lower piece. The location of the nozzle in relation to the joint should be as shown in figure 65-5.

4. Weld the joint. Allow for distortion by running the first bead on the opposite side from the tacks.

5. Cool and examine the bead for uniformity. The weld metal should be equally distributed between both pieces and show no signs of undercut.

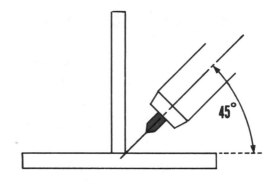

Fig. 65-5 Transverse angle

Note: A tack weld should be strong enough to resist cracking during the welding process but not large enough to affect the appearance of the finished weld. This is done in MIG welding by using slightly higher wire-feed speed.

6. Make another T joint welding only the tack side. Test this weld by bending the top piece against the joint a full 90 degrees. Examine the joint for root penetration and uniform fusion.

7. Continue to make the fillet weld until acceptable welds can be made each time. The fillet-type weld used on the lap and T joint is the most common weld.

Welding Procedure (Corner Joint)

Note: The technique used to set up and align the pieces to be joined for the down-hand corner joint is more difficult. Figure 65-6 shows how this is done without using a fixture.

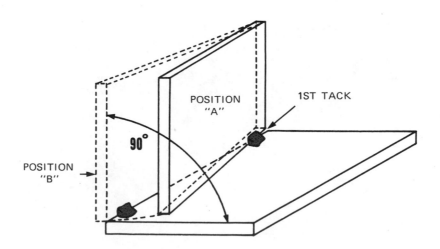

Fig. 65-6 Setting up for corner joint

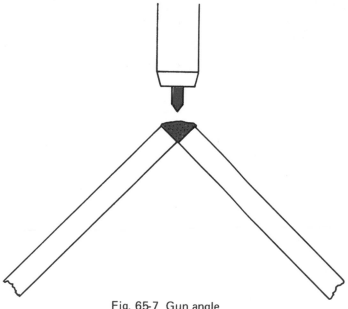

Fig. 65-7 Gun angle

1. Make the first tack while holding the piece as shown by position A, Figure 22-6. Then lift the hood and align the top piece to position B. Make a perfect open corner joint before placement of the second tack.

2. Maintain a stickout of 1/4 inch to 3/8 inch.

3. The longitudinal angle is 10 degrees from perpendicular using a trailing gun angle. The transverse gun angle should be perpendicular or bisect the included angle. See figure 65-7.

4. Weld the joint. Pay close attention to the start and end of the joint to avoid buildup or washout.

5. Cool and examine the bead for uniformity and penetration. The weld metal should be equally distributed between both pieces and show no signs of undercut or overlap. See figure 65-8.

Fig. 65-8 Examples of uniform welds

6. To test the corner joint, place the welded unit on an anvil and hammer it flat in order to examine root fusion and penetration.

7. Make more corner joints until they have uniform appearance and a good finish contour. The opposite side of this joint provides for good fillet weld practice.

Checking Application

1. Recheck the equipment settings according to the operation manual.

2. Keep the equipment clean.

3. Practice these four basic joints in the flat position using thicknesses of material up through 3/16 inch.

REVIEW QUESTIONS

1. When a tack is placed on two pieces of material being joined, what is the function of the tack?

2. What are the two types of nozzle angles as related to the longitudinal angle used in the operation of the gun?

3. What causes undercut and why is it harmful to the strength of the weld?

4. When welding any joint, what is important concerning bead location?

5. Of the four joints, butt, lap, corner and tee, which one might require more inert gas? Why?

UNIT 66   PROCEDURE VARIABLES

OUT-OF-POSITION WELDING

Upon satisfactory completion of the welds in the flat position, the student will be able to use the acquired skill and knowledge to weld out of position. This includes horizontal, vertical-up, vertical-down, and overhead welds. The basic procedures for each individual joint are no different out of position than in the flat position except a reduction in amperage of 10 percent is usually recommended. See chart 63-2.

MISALIGNED MATERIALS

The operator may, at times, have to weld pieces of material that are not in plane or aligned properly. There may be gaps or voids of various sizes which need a variation of stickout and/or wire-feed speed and voltage. The student should practice on the joints that require a deviation from standard procedures.

WELDING HEAVIER THICKNESSES

Heavier thicknesses of material can be welded with the MIG process using the multipass technique. This is done by overlapping single small beads or progressively making larger beads, using the weave technique, as in figure 66-1. The numbers refer to the order in which the passes are made. Individual job requirements govern the end result.

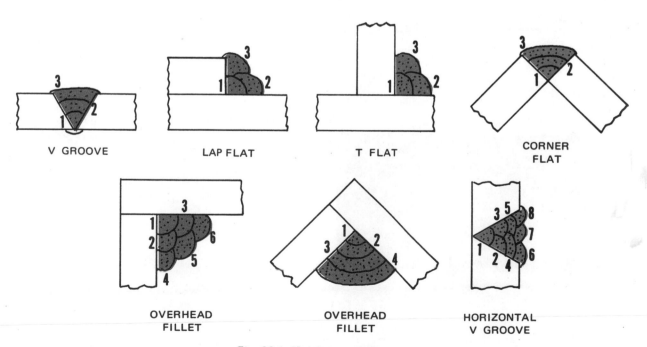

Fig. 66-1 Multipass welding

DEFECTS	PROBABLE CAUSE	CORRECTIVE ACTION

Fig. 66-2 Porosity

Gas flow does not displace air, clogged or defective system, frozen regulator

Set gas flow between 15 and 23 CFH. Clean spatter from nozzle often. Use a regulator heater when drawing over 25 CFH of CO_2

Fig. 66-3 Porosity in crater at end of weld

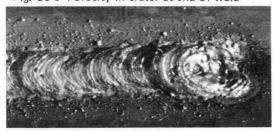

Pulling gun and gas shield away before crater has solidified

Reduce travel speed at end of joint

Fig. 66-4 Cold lap lack of fusion

Improper technique preventing arc from melting base metal

Direct the welding arc so that it covers all areas of the joint. Do not allow the puddle to do the fusing. Use a slight whip motion.

Fig. 66-5 Burn-through and too much penetration

Heat input too high in the weld area

Reduce wire-feed speed to obtain lower amperage. Increase travel speed. Oscillate gun slightly. Increase stickout to 1/2 inch maximum

Fig. 66-6 Lack of penetration

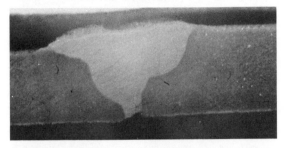

Heat input too low in the weld area

Increase wire-feed speed to obtain higher amperage. Reduce stickout to 1/4 inch.

Fig. 66-7 Whiskers

Electrode wire pushed past the front of the weld puddle leaving unmelted wire on the root side of the joint

Cut off ball on end of wire with pliers before pulling trigger. Reduce travel speed and, if necessary, use a whipping motion.

Fig. 66-8 Wagon tracks

Too high bead contour or too high crown. Area where bead fuses to side of joint is depressed and next bead may not completely fill depressed area or void.

Arc voltage and travel speed should be high enough to prevent crown. When welding over these areas, be sure that the welding arc melts the underlying weld and base metal.

CAUSE AND CORRECTION OF DEFECTS

The operator has to be able to recognize and correct possible welding defects. MIG welding, like the other processes, must be properly applied and controlled to consistently give high quality welds. The defects are shown in figures 66-2 through 66-8, accompanied by the causes and corrective actions to be taken.

REVIEW QUESTIONS

1. What positions does out-of-position welding refer to?

2. What is porosity in a MIG weld?

3. What quality will the weld probably lack if the current input is too low at the arc?

4. In MIG welding heavy plate, is more amperage and voltage required than welding light plate? Why?

5. Can the MIG gun be oscillated to improve bead conformity?

SECTION 4
Metallurgy and the
Welding Industry

Metallurgy is concerned with the application of chemistry and physics to metals and their processing. Those concerned with the complex changes in metals as they are heated and cooled in the welding process must be familiar with the metallurgy of a large variety of metals and alloys. However, knowledge of the characteristics, identification, and uses of iron and steel is the most important.

Since welding operations are always concerned with the strength of metals, a knowledge of the testing of metals becomes essential to all personnel involved in welding. The physical characteristics of metals and the effects of heating and cooling can be illustrated by some relatively simple experiments.

UNIT 67   METALS AND ALLOYS

THE FIELD OF METALLURGY

Metallurgy has been described as an exact science which uses other sciences. A metallurgist has a broad knowledge of chemistry and physics. Colleges and universities grant degrees in metallurgy. Producers of metals employ teams of scientists, metallurgical engineers, and technicians to ensure a high-quality product, and to assist their customers in obtaining the best and most efficient use of the product. These teams of experts also develop new alloys and improve the characteristics of present metals and alloys.

ALLOYS

Cast iron and steel are the materials most often encountered by the welder. Therefore, the major emphasis in this unit will be on the various characteristics of iron and the ferrous alloys as they apply to the field of welding. An *alloy* is a mixture of two or more elements, at least one of which is a metal, which may have different characteristics than either of the base elements. Alloys are divided into two general groups — ferrous alloys and nonferrous alloys. A *ferrous alloy* is one in which the major alloying element is iron. The *nonferrous alloys* contain little or no iron. They are a mixture of various elemental metals as shown in figure 67-1.

Elemental metals are very seldom found in a free state in nature; they are usually in the form of ores which require considerable refinement in order to obtain a pure metal.

ORE TO IRON

The refining of iron from its ore, generally FeO or Fe_2O_3, is not difficult, but it requires the use of very large equipment to be economically practical.

Figure 67-2 shows the materials and steps necessary to produce the basic pig iron, the further reduction for steel, as well as the rolling mill process used to produce the basic steel shapes — blooms, slabs, and billets. Below this is a brief explanation of the principal products rolled or drawn from each of the basic shapes.

To produce 1 ton of pig iron, it is necessary to charge the blast furnace with about 1 3/4 tons of iron ore, 3/4 ton of coke, and 1/3 ton of limestone as well as 4 tons of air

ALLOY	ELEMENTAL METALS
Brass	Copper-Zinc
Bronze	Copper-Tin
Solder	Tin-Lead
Dural	Aluminum-Copper
White Metal	Aluminum-Zinc

Fig. 67-1 Common nonferrous alloys

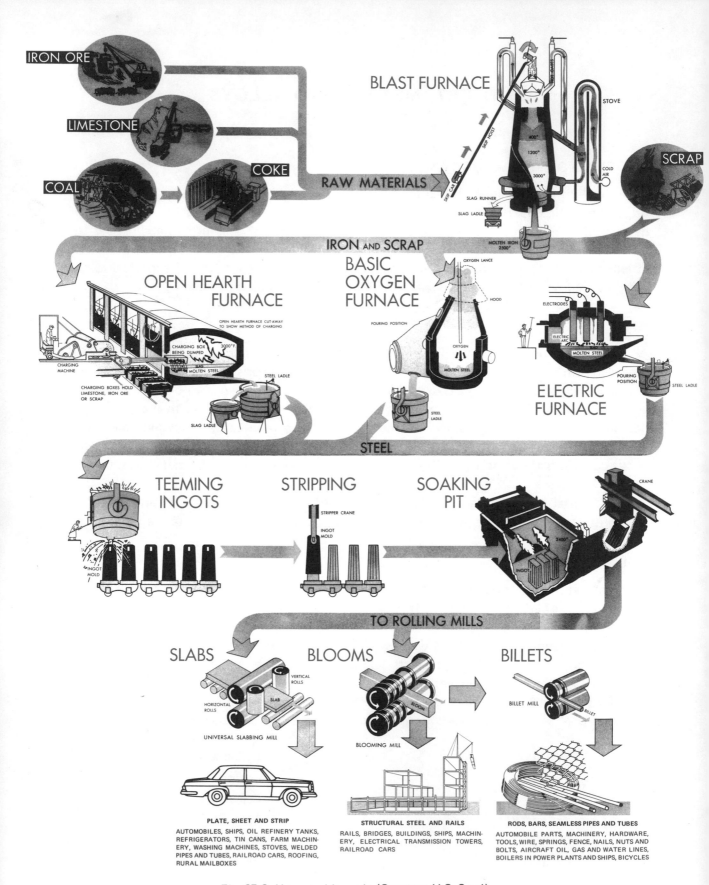

IRON ORE

LIMESTONE

COAL

COKE

RAW MATERIALS

BLAST FURNACE

STOVE

SCRAP

SKIP HOIST

600°
1200°
3000°

HOT

COLD AIR

SLAG RUNNER

SLAG LADLE

MOLTEN IRON 2500°

IRON AND SCRAP

OPEN HEARTH FURNACE

OPEN HEARTH FURNACE CUT-AWAY TO SHOW METHOD OF CHARGING

CHARGING BOX BEING DUMPED

3000°F

SLAG
MOLTEN STEEL

CHARGING MACHINE

CHARGING BOXES HOLD LIMESTONE, IRON ORE OR SCRAP

STEEL LADLE

SLAG LADLE

BASIC OXYGEN FURNACE

OXYGEN LANCE

HOOD

POURING POSITION

OXYGEN

MOLTEN STEEL

STEEL LADLE

ELECTRIC FURNACE

ELECTRODES

ELECTRIC ARC

MOLTEN STEEL

POURING POSITION

STEEL LADLE

STEEL

TEEMING INGOTS

INGOT MOLD

STRIPPING

STRIPPER CRANE

INGOT MOLD

SOAKING PIT

CRANE

2400°

INGOT

TO ROLLING MILLS

SLABS

VERTICAL ROLLS

HORIZONTAL ROLLS

SLAB

UNIVERSAL SLABBING MILL

BLOOMS

BLOOM

BLOOMING MILL

BILLETS

BILLET MILL

BILLET

PLATE, SHEET AND STRIP

AUTOMOBILES, SHIPS, OIL REFINERY TANKS, REFRIGERATORS, TIN CANS, FARM MACHINERY, WASHING MACHINES, STOVES, WELDED PIPES AND TUBES, RAILROAD CARS, ROOFING, RURAL MAILBOXES

STRUCTURAL STEEL AND RAILS

RAILS, BRIDGES, BUILDINGS, SHIPS, MACHINERY, ELECTRICAL TRANSMISSION TOWERS, RAILROAD CARS

RODS, BARS, SEAMLESS PIPES AND TUBES

AUTOMOBILE PARTS, MACHINERY, HARDWARE, TOOLS, WIRE, SPRINGS, FENCE, NAILS, NUTS AND BOLTS, AIRCRAFT OIL, GAS AND WATER LINES, BOILERS IN POWER PLANTS AND SHIPS, BICYCLES

Fig. 67-2 How steel is made (Courtesy U.S. Steel)

heated to about 1,200 degrees F. (649 degrees C). (Twelve and one-half cubic feet of air weighs one pound.)

The *blast furnace* is actually a huge chemical distillery, 100 feet tall, 30 feet in diameter at the bottom, and tapering to a much smaller diameter at the top. It is lined with heat-resistant brick. Once put in operation, this type of furnace is continually recharging until it breaks down and needs repairs or replacement.

The heated air forced under pressure into the bottom of the furnace combines with the coke, producing heat (3,000 degrees F./1,649 degrees C) and carbon monoxide gas (CO). Hot carbon monoxide combines chemically with the oxygen in the hot iron ore (FeO) to form carbon dioxide (CO_2) which leaves the furnace at the top and is piped through the stoves that heat the incoming air. This conserves some of the heat that would be wasted otherwise. This chemical reaction causes the iron to combine with the carbon in the coke and other impurities. At this point, the iron and iron-carbon compound mixed with the impurities becomes hot enough to melt and flow to the bottom of the furnace. The limestone also melts and combines with some of the earthy impurities of the ore to form a slag which floats on top of the molten iron.

Every 4 to 6 hours the iron is drawn from the furnace into refractory-lined ladles or ladle cars. It is then cast either into *pigs* (small blocks) for use by iron foundries which make cast-iron products, or it is poured into either open-hearth furnaces or Bessemer converters for further refinement into steel. The slag is removed from the top of the iron at frequent intervals.

REFINEMENT OF IRON

The reduction of pig iron into steel is necessary due to the impurities in the iron. There is very little refinement in the blast furnace other than the reduction of the oxygen content of the ore. Certain impurities, such as sulphur and phosphorus, remain. At high temperatures, the iron tends to pick up carbon from the coke. Pig iron contains from 2 to 4.5 percent carbon. This is far above the acceptable carbon content for steel which usually ranges from 0.04 to 1.7 percent.

The bulk of plain carbon steels is produced by the *open-hearth* process. The furnace is charged with pig iron and scrap steel. The impurities are reduced by the use of special linings in the furnace plus the addition of chemical elements to the molten material. Each batch, produces from 50 to 500 tons of plain-carbon steel.

The Bessemer converter is charged with 25 to 50 tons of molten iron and a blast of air is blown through the metal to burn off the impurities. In either process the molten metal is drawn off into ladles or ladle cars and poured into *ingot* molds. These ingots are the basic building blocks of the steel industry. They are put through a series of rolling and drawing processes to produce mammoth structural shapes and all lesser shapes and sizes down to wires less than the diameter of a human hair.

THE ELECTRIC FURNACE PROCESS

Many of the special carbon steels and the alloy steels are produced by the electric furnace process. The furnace is charged with scrap iron and alloy steel in the proper proportions. These materials are then melted by an electric arc passing between large carbon or graphite electrodes. The molten steel is constantly checked by the metallurgical laboratory, and alloying elements are added in the proper amounts. Other materials are also added that act as a flux to remove unwanted elements. The metals or alloys produced by the electric furnace process have a chemical content that is very closely controlled.

Most of the rod and electrodes used in the welding industry are the product of electric furnaces. This is primarily due to the necessity of maintaining a very low sulphur and phosphorus content in the rods if a weld of superior quality is to be produced.

REVIEW QUESTIONS

1. Is the metallurgist necessarily a skilled mathematician?

2. For what reasons do you think metals are alloyed rather than being used in their original state?

3. Explain the difference between a ferrous and nonferrous metal?

4. What are the proportions of a charge that goes into a blast furnace?

5. The text indicates that this charge will produce one ton of pig iron. Can this vary?

6. Most iron ore is shipped for considerable distances to the blast furnace site for refining. Why is it not refined at the mines?

7. An oxyacetylene flame with an excess of acetylene is often referred to as a reducing flame. Draw a parallel between this flame and the action in a blast furnace.

8. What are the distinguishing differences and similarities between blooms, slabs, and billets?

9. How many cubic feet of air are generally required to produce 1 ton of iron?

10. In making consumer products, why is the iron from the blast furnace cast, rather than rolled or drawn?

11. The text indicates that certain impurities are burned out of the molten pig iron to produce steel. If an oxidizing flame is used to weld steel, the metal is burned and an unacceptable weld is produced. Why does the same condition not exist when producing steel?

12. The flow chart, figure 67-2, indicates that scrap metal is used in all the steelmaking processes. What problems does this make for steel makers?

UNIT 68   CHARACTERISTICS OF STEEL

PHYSICAL CHARACTERISTICS

All personnel engaged in the fabrication of welded products are concerned to some degree with the chemical content and physical characteristics of the materials being used. The material may be sheared, sawed, punched, heated (rapidly or slowly), hot worked, cold worked, drilled, bent, twisted, and otherwise treated. The personnel are further concerned with the strength, yield point, ductility, fatigue value, impact value, hardness, conductivity (electrical and thermal), thermal expansion, the effect of the rate of heating and cooling, the effect of various alloying elements on the melting point, and the effect of various impurities on the weld and adjacent zone.

In some instances, it is necessary to consider heat-treating the finished product to reduce internal stresses or to alter the grain structure of the material in and near the welding zone.

IDENTIFYING STEELS

All producers and users of steel and its alloys are concerned with identifying the material according to its chemical content.

American mass production methods in the metal industries proceed on the basis that each manufacturer can have the exact metal that any product requires. Metals must be of uniform quality in every shipment. For this reason, standards and numbering systems for identifying carbon steels and alloy steels have been set up by the SAE (Society of Automotive Engineers) and the AISI (American Iron and Steel Institute). The two systems are very similar. The main difference is that in the AISI system, a letter indicating the steelmaking process precedes the number. The use of this letter is important because two steels having practically the same makeup but made by different processes will often have slight but important differences in their properties. These AISI letter prefixes, with their meanings, are as follows:

A   Open-hearth alloy steels
B   Acid Bessemer carbon steel
C   Basic open-hearth carbon steel
D   Acid open-hearth carbon steel
E   Electric furnace steel

In the SAE numbering system, an index system of numbers is used to identify the composition of the various steels. These sets of numbers are very convenient for placement on drawings and blueprints to specify the types of steel to be used.

Each number consists of four (or five) digits, figure 68-1. The *first digit* at the left indicates the type of steel. The *second digit* indicates either the average percent of the main alloying element (other than carbon) or the presence of a second alloying element. The *last two or three digits* indicate the average carbon content in points (A point is .01 percent.)

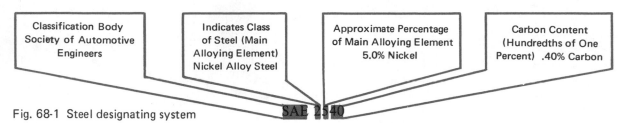

Fig. 68-1 Steel designating system

Sometimes the prefixes TS and X are used. The prefix X denotes variations in the percent range of elements. The prefix TS denotes tentative standard; generally, a new composition has not yet been adopted completely.

In the list of steel types in figure 68-2, the small x's at the end of the numerals indicate the places of digits. These digits, when put in place, further subdivide each type of steel into subtypes as mentioned. Carbon steels are often referred to as *low-, medium-,* or *high-carbon steels.*

It must always be kept in mind that the physical characteristics and chemical content of the metals have a very definite relationship. Some individual subtypes of plain-carbon steels (classification 10xx), with the percent ranges of their carbon content, are given in figure

TYPE OF STEEL	SAE NUMERALS
Carbon Steels	1xxx
Plain carbon	10xx
Free cutting (screw stock)	11xx
High manganese	13xx
Nickel Steels	2xxx
3.50% nickel	23xx
5.00% nickel	25xx
Nickel Chromium Steel	3xxx
1.25% nickel, 0.60% chromium	31xx
3.50% nickel, 1.50% chromium	33xx
Molybdenum Steels (0.25% molybdenum)	4xxx
Chromium 1.0%	41xx
Chromium 0.5%, nickel 1.8%	43xx
Nickel 2%	46xx
Nickel 3.5%	48xx
Chromium Steels	5xxx
Low chrome	51xx
Medium chrome	52xx
Chromium Vanadium Steels	6xxx
Nickel-Chromium-Molybdenum (low amounts)	8xxx
Silicon-Manganese	92xx

Fig. 68-2 Basic SAE numbering system for steels

SAE NUMBER	CARBON RANGE PERCENT	SAE NUMBER	CARBON RANGE PERCENT	SAE NUMBER	CARBON RANGE PERCENT
1010	.08 – .13	1040	.37 – .44	1070	.65 – .75
1015	.13 – .18	1045	.43 – .50	1075	.70 – .80
1020	.18 – .23	1050	.48 – .55	1080	.75 – .88
1025	.22 – .28	1055	.50 – .60	1085	.80 – .93
1030	.28 – .34	1060	.55 – .65	1090	.85 – .98
1035	.32 – .38	1065	.60 – .70	1095	.90 – 1.03

Fig. 68-3 SAE specifications for plain carbon steels

SAE NUMBER	CARBON RANGE PERCENT	SAE NUMBER	CARBON RANGE PERCENT	SAE NUMBER	CARBON RANGE PERCENT
1112	0.13 max.	1115	.13 – .18	1120	.18 – .23

Fig. 68-4 SAE specifications for free-cutting carbon steels

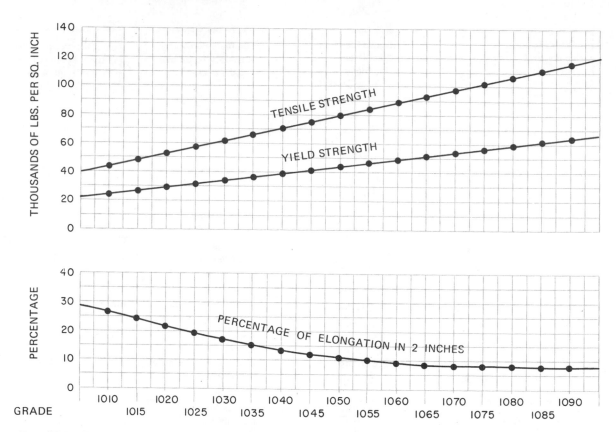

Fig. 68-5 Effect of carbon content on physical characteristics and average minimum properties for hot-rolled steel

68-3. Some subtypes of free-cutting carbon steels are shown in figure 68-4. Figure 68-5 will be referred to many times in the discussion of the physical characteristics of steel.

REVIEW QUESTIONS

1. Define the term physical characteristics.

2. Under the AISI steel identifying system, what prefix letter would be used if the material was to be used for welding rods?

3. Using the basic SAE chart, figure 68-2, how is the percentage of nickel content determined?

4. Using the same SAE system, what distinguishes the steel alloys chrome-molybdenum and nickel-molybdenum from each other?

5. SAE 4130 is a popular high-strength steel used in the fabrication of aircraft frames. From the information available in figure 68-2, what is the formula for this material?

UNIT 69   DESTRUCTIVE TESTING OF STEEL

TENSILE STRENGTH

One of the prime reasons for using steel or its alloys as a material of construction is its strength. *Tensile strength* or *ultimate tensile strength* is a measure of the material's resistance to being pulled apart. The standard of measurement for tensile strength in materials such as steel is the number of thousands of pounds of load required to pull apart a bar 1 inch square. To put it another way, tensile strength is the ratio of load to original cross-sectional area.

Size of Specimens

While there are tensile-testing machines capable of pulling apart specimens of 1-inch cross-sectional areas, the usual procedure is to test a specimen of a considerably smaller cross section. One of the most popular of these specimens is shown in figure 69-1.

The center section of this specimen is turned to a diameter of .505 inch, which makes the area of the end of the cross section one-fifth of a square inch and aids rapid calculation. Since most tensile-testing machines have a direct reading dial which indicates the load in pounds, the technician or engineer has only to multiply the dial reading by five in order to determine the tensile strength. Example: Dial reading of 12,000 lbs. x 5 = 60,000 psi. (pounds per square inch) tensile strength.

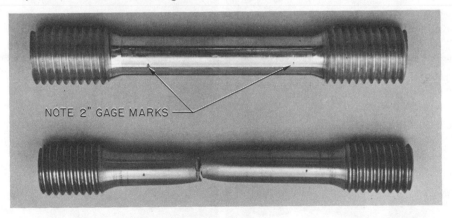

NOTE 2" GAGE MARKS

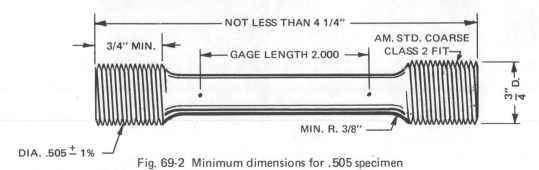

NOT LESS THAN 4 1/4"

3/4" MIN.

GAGE LENGTH 2.000

AM. STD. COARSE
CLASS 2 FIT

3"/4 D.

MIN. R. 3/8"

DIA. .505 ± 1%

Fig. 69-2 Minimum dimensions for .505 specimen

The generally accepted dimensions for a .505 specimen are shown in figure 69-2.

In certain instances the material being tested may not be thick enough to produce a .505 specimen. In these cases, a ratio of the dial reading in pounds to the cross-sectional area in square inches can be adopted. For example, if a plate yields a 3/16 inch x 1 inch cross-sectional specimen and the dial reading is 10,000 pounds, the following calculation is made: .1875 x 1 = .1875 square inch. Therefore, the ratio would be

$$\frac{10{,}000 \text{ pounds}}{.1875 \text{ sq. in.}}$$

which would equal 53,333 p.s.i. tensile strength.

Tensile Testing

Each step in the tensile-testing procedure must be carried out with careful attention to accuracy. All dimensions should be taken with a reliable micrometer and carefully recorded.

Most tensile-testing machines, such as the one shown in figure 69-3, are usually operated by hydraulic pressure, but some are mechanically operated. Such machines generally have a direct-reading dial indicating the total load being applied. These dials are usually equipped with two pointers. One pointer indicates the load at any given instant, and the other pointer, a friction type activated by the original pointer, remains at the highest point

Fig. 69-3A Mounting tension specimen in tension machine

Fig. 69-3B Testing the specimen

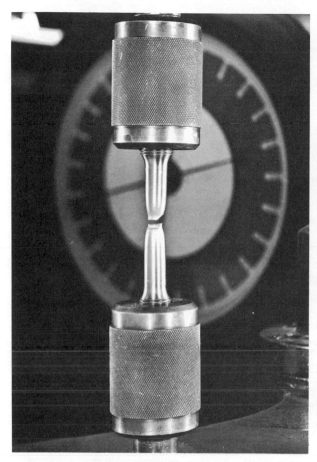

Fig. 69-3C Broken specimen

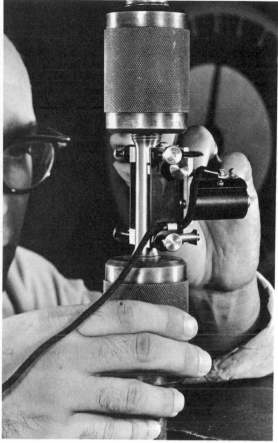

Fig. 69-3D Attaching extensometer to specimen

reached on the dial until it is reset manually by the operator. This ensures an accurate reading of the maximum load applied.

Figure 69-3D illustrates the use of an *extensometer.* An extensometer is a micrometer device used during the tensile testing procedure. Measurements are taken prior to and after the testing process. The results are recorded and used to determine the amount of stretch a specimen will tolerate before it pulls apart.

Tensile-testing machines are adjustable for different sizes and shapes of specimens and are generally designed so that compression tests can also be made on them. As a general rule, the *compression strength* of carbon steel is considered to be three-fourths of the tensile strength. For example a steel with a tensile strength of 60,000 p.s.i. would have a compression strength of about 45,000 p.s.i. This rule does not hold for all materials, even in the ferrous metals. For instance, grey cast iron has a tensile strength in the 40,000 to 45,000 p.s.i. range, while its compression strength is about 90,000 p.s.i. This makes it an excellent material for bearing loads which are strictly compressive.

YIELD POINT

While tensile tests are being made, other pertinent information is usually obtained and recorded. The *yield point,* also called *yield strength,* is the point at which the specimen will continue to stretch to some extent even though no additional load is applied.

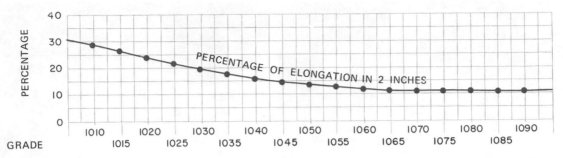

Fig. 69-4 Effect of carbon content on physical characteristics and average minimum properties for hot-rolled steel

ELASTIC LIMIT

The *elastic limit* is usually described as the point reached in stressing a material beyond which a further stress produces a permanent deformation. Steel is perfectly elastic up to its elastic limit. If a load of 4,000 p.s.i. is applied to a specimen, the distance between 2-inch gage marks will increase. If the load is released, the material will return to its original dimensions. If the load is doubled to 8,000 p.s.i., the difference in length, or amount of stretch, will be double that at 4,000 p.s.i. This ratio of stretch to stress will remain consistent all the way up to the elastic limit. Beyond this point the material never returns to its original dimensions.

If the cross-sectional area is checked with a micrometer during the testing, it is found that the diameter becomes smaller as the stress is increased. Once the elastic limit is reached, the specimen remains smaller in diameter when stress is released.

This procedure of exceeding the elastic limit is encountered daily in numerous fabricating plants in such instances as rolling cylinders in plate and sheet rolls, making stampings in mechanical or hydraulic presses, using forming dies, and twisting metals for ornamental iron work.

ELONGATION

Refer to figure 69-4. Observe that the vertical, or left-hand, axis indicates percentage, but the horizontal axis still indicates the grade or percentage of carbon. This curve goes lower rather than higher as the carbon content increases. These values are also found during the process of determining the tensile strength and yield point.

In the testing procedure, enough stress is put on the specimen to destroy it and then the broken ends are brought together until they touch and exactly match. Measure the distance between the gage marks. The difference between the original gage length (2 inches) and the final gage length divided by the original gage length equals the *elongation in percentage points.*

When L equals the original length, and L_1 equals the final length, then

$$\frac{(L_1 - L)}{L} \times 100 = \text{percentage of elongation}$$

For example: If the final gage length is 2 1/2 inches, then

$$\frac{2\ 1/2\ \text{in.} - 2}{2} \times 100 = \frac{.50}{2.0} \times 100 = 25\% \text{ elongation}$$

REDUCTION IN AREA

The measure of *reduction in area* is also obtained from the original test specimen. After the sample has been broken, the smallest diameter of the necked-down portion is measured. The decrease of this cross-sectional area is expressed as a percentage of the original area.

If A is the original cross-sectional area and A_1 is the smallest cross-sectional area after rupture, then

$$\frac{(A - A_1)}{A} \times 100 = \text{percentage reduction in area.}$$

Example: Original diameter .505,

$$\text{Original area } \left(\frac{.505}{2}\right)^2 \times 3.1416 = .2002 \text{ sq. in.}$$

$$\text{Smaller diameter } A_1 = \left(\frac{.357}{2}\right)^2 \times 3.1416 = .1001 \text{ sq. in.}$$

$$\text{Then substituting for } \frac{A - A_1}{A} \times 100 \text{ percent}$$

$$\frac{.2002 - .1001}{.2002} \times 100 = \frac{.1001}{.2002} \times 100 = 50 \text{ percent}$$

METAL FATIGUE

Another physical characteristic of a metal which is important to the welding industry is the *fatigue limit,* also called the fatigue value or endurance limit. *Fatigue* is the fracture (or break) of a material under repeated varying stresses which have a maximum value that is less than the tensile strength of the material.

Fatigue tests are performed on several highly polished specimens. These are placed in the chucks of a machine, as shown in figure 69-5.

The first specimen is stressed to a point beyond its estimated fatigue value or endurance limit. The machine is turned on and, as it rotates, the specimen is alternately bent twice for each revolution. In this case, failure is usually rapid. Additional specimens are tested by reducing the stress in the outer crystals each time until a point is reached at which the specimen can withstand 10 million reversals with no evidence of failure. This point is then recorded and reported as the fatigue value in pounds-per-square-inch stress in the outer crystals.

The fatigue value of steel is usually between 45 and 55 percent of its ultimate tensile strength. Repeated alternate stresses beyond this proportion will usually cause ultimate

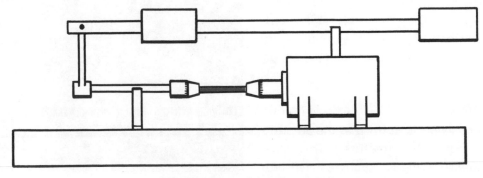

Fig. 69-5 Setup for fatigue testing

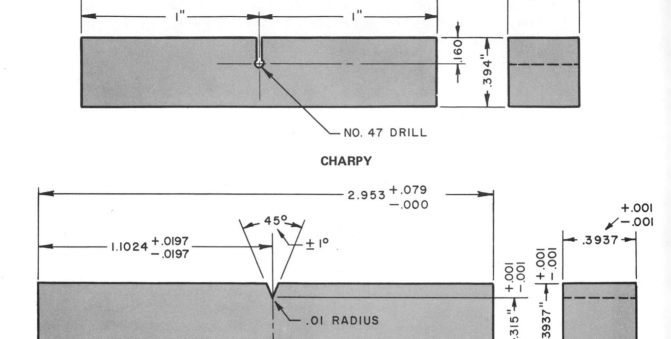

CHARPY

IZOD

Fig. 69-6 Dimensions for impact specimens

failure.   An excellent example of fatigue resistance is the automobile common leaf spring which is flexed millions of times during its lifetime.

One of the common false statements heard is, "the metal crystallized and broke." Actually, metal is always crystalline in structure except when molten. When metal is alternately stressed beyond its endurance limit, the crystals do not break, but slip past each other. Thus, when fatigue occurs, these crystals stand out more prominently than they would in other types of failure such as tension or impact.

IMPACT VALUE

Impact value is defined as the amount of energy required to fracture a test specimen. Impact- testing is usually carried out using either the *Izod* or *Charpy* testing

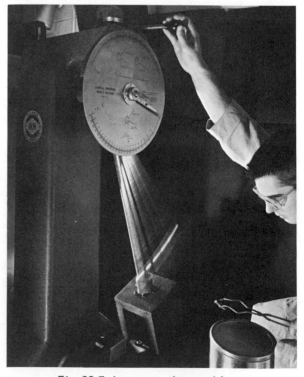

Fig. 69-7 Impact-testing machine

methods. The Charpy and Izod testing procedures differ in the ways in which the specimens are prepared, figure 69-6. The specimen is held by both ends in the impact-testing machine in the Charpy procedure and by only one end in the Izod procedure.

In both tests the specimens are given a fracturing blow produced by a pendulum, always of the same weight, that swings through a measured arc, always of the same length, figure 69-7. Metals of high impact value absorb most of the impulse energy of the swinging weight at the point of impact and the pendulum continues its swing through a relatively short arc.

In metals of lower impact value, less energy is absorbed by the test specimen and the pendulum swings through a longer arc. The impact-testing machine is fitted with a scale to indicate the length of the arc, and the results are reported in foot-pounds of energy absorbed by the specimen.

Large establishments usually have an engineering department that controls the specifications, such as the types of welding rods used and the procedures to be followed. In job shops and smaller establishments, the choice of rods is left up to the welder.

Producers of welding rods will supply information on all the physical characteristics of the weld metal produced by their electrodes on request. The catalogs of their welding rods usually contain most of the information along with a description of the uses and procedures for each type of rod.

The welder should be aware that there is a great difference in the impact value or impact strength of the welds produced by different welding rods. For instance, electrodes in the E-0012 and E-6013 series produce welds of excellent strength. A tensile strength of 70,000 to 78,000 p.s.i. is not unusual. The impact value is usually in the 25 to 30 foot-pounds range (Charpy). If the load is applied gradually, these welds will stand up as well as, or better than, the material being welded.

If, however, the load is applied in the form of repeated sudden shocks, such as the impact a trailer hitch undergoes on a rough road, then the welds produced by the E-6010 or E-6011 electrodes will be much better. These types of electrodes do not produce welds quite as high in tensile strength as the E-6012-13 series (usually in the 60,000 to 65,000 p.s.i. range), but the impact value is in the 50- to 70-foot-pound range (Charpy).

Actually the welds in the 60,000 to 65,000 p.s.i. range are as good as, or better than, most commercial grades of structural or plate steel.

EXPERIMENT: TESTING ELASTIC LIMIT

Materials

Straight steel gas welding rod, 1/8 in. x 36 in.
Gray cast iron gas welding rod, 1/8 in. x 36 in.

Procedure

1. Grasp a welding rod at each end and bend it until it forms an arc with a fairly large radius. Release the pressure and note that the rod returns to its original straight position.

2. Continue to bend the rod several times, increasing the pressure a little at a time.

   Note: A point is reached when the rod no longer returns to its straight position and remains bent. This is actually a combination of tension and compression. The crystals of metal on the outer side of the curved rod are in tension, and the crystals on the inner side are in compression.

3. Repeat the procedure with a gray cast iron welding rod.

   CAUTION: Be sure that the cast iron rod is bent *down and away* from you.

Observation and Conclusion

Observe the results and compare the correlation between the yield strength and the ultimate tensile strength of gray cast iron and carbon steel.

REVIEW QUESTIONS

1. Define the term, *ultimate tensile strength.*

2. How is the tensile strength of a material determined?

3. Name a common product that utilizes tension almost exclusively in order to be effective and useful?

4. What is the difference between exceeding the elastic limit of a material and reaching or exceeding the yield point?

5. Explain the difference between ductility and percentage of elongation.

6. Explain the relationship between ductility and the reduction in area.

7. When is a metal not crystalline in structure?

8. In general, would you expect that steels of high- or low-carbon content would have the higher impact value?

UNIT 70   NONDESTRUCTIVE TESTING OF STEEL

Notice that the tests previously explained are destructive in nature, that is the specimen is broken or destroyed in the process of making the tests. This unit will explain many nondestructive tests used to determine other physical characteristics of metals important to welders.

HARDNESS

One of the more important and commonly observed characteristics is *hardness.* Hardness can be described as the resistance to penetration of a particular material. The hardness of steel may be increased through the addition of carbon to the material.

The ordinary knife or razor blade would be of little use if it were made of soft steel. The bearings, valves, and springs of our automobiles would have a very short life if they were not hardened to some degree.

One of the original systems devised to indicate hardness was Mohs' scale — a 10-point scale of minerals arranged in an ascending scale from talc (1) to diamond (10). This system would be considered crude by today's standards.

Another method, not generally used in today's testing laboratories, is the *scleroscope,* or Shore hardness tester. In this test, a calibrated glass tube is fitted with a hardened weight or a weight with a diamond tip, which is dropped from a measured height. The weight bounces, and the height of the bounce is observed and recorded as the Shore hardness number on the Shore scale.

One of the more interesting adaptations of the Shore test is that used to inspect many of the millions of hardened steel balls used in the manufacture of ball bearings, figure 70-1.

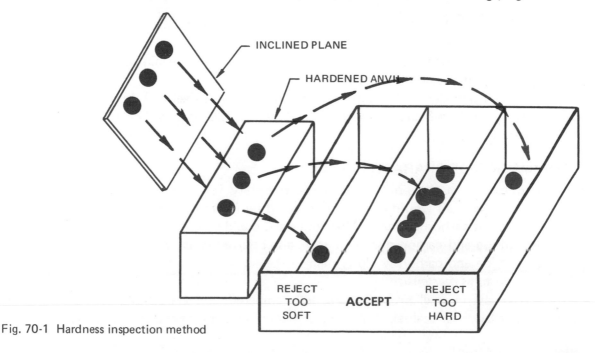

Fig. 70-1 Hardness inspection method

Fig. 70-2 Rockwell hardness tester

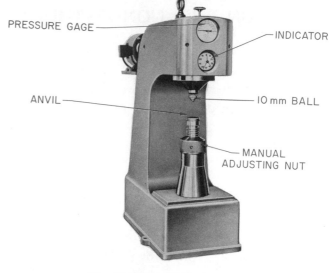

Fig. 70-3 Brinell hardness tester

The two most commonly used machines for testing hardness are the *Brinell* and *Rockwell* hardness testers figures 70-2 and 70-3. In both systems, pressure on the penetrator forces it into the test piece, and the amount of penetration is measured and recorded as the hardness number.

In the Brinell hardness tester, a hardened steel ball, 1/8 inch in diameter, is used as the penetrator. The amount of penetration is measured with a special calibrated microscope which measures the diameter of the indentation in millimeters. Some types of Brinell testers are equipped with a direct-reading dial for convenience and speed.

The Rockwell hardness test uses a hardened steel ball, 1/16 inch in diameter, for measuring the softer metals, and the results are recorded as the hardness number on the Rockwell B-scale.

For harder metals the steel ball is replaced by a conical diamond point commonly referred to as a *diamond brale.* As with the 1/16-inch ball, the amount of penetration of the diamond brale into the test piece is measured on a direct-reading dial. Results are recorded as the hardness number on the Rockwell C-scale.

There is a definite relation between the hardness of a steel and its tensile strength. If the Brinell number in figure 70-4 is multiplied by 500, the result will be approximate to the tensile strength.

COMBINATION OF CHARACTERISTICS

Hardness is not the only characteristic necessary for carbon and alloy steel. If this were the case, all that would be necessary would be for the steelmakers to produce a steel with a sufficient carbon content. Then the users could heat it to the temperature at which the iron carbide and carbon dissolve in iron, also referred to as *austenite.* By quick cooling, the iron carbide and carbon would remain in the form of an extremely hard chemical mixture, referred to as *martensite.*

A high-carbon steel properly hardened, tempered, and sharpened would make an excellent razor blade or scalpel. However, a cold chisel made of the same material and heat-treated in the same manner would shatter when subjected to hammer blows. The chisel steel must be made of an alloy that maintains a reasonably sharp cutting edge and is sufficently tough to withstand impact.

The teeth on draglines and power shovel buckets are subjected to extreme abrasion in addition to the severe impact produced by forcing the bucket into the rock. A desirable material in this case is one that has an extremely hard abrasion-resistant exterior and a tough impact-resistant interior. Shovel teeth and buckets are often made of a steel with the principal alloying element of manganese in sufficient quantity to cause the alloy to be self-hardening. As the manganese steel is subjected to constant impact as it is used, the surface *work-hardens.* As abrasion wears off this surface, the surface directly underneath work-hardens. The process repeats itself until the tooth or bucket is worn beyond its useful stage.

Cutting and forming tools with fine cutting edges, that are needed in the machine trades, can be produced by hardening and tempering plain carbon steels. However, today's high-speed production methods require rather complicated alloys to produce tools that will give long service under the excessive heat generated in the work and tool.

Most metals work-harden to some extent. Even low-carbon, cold-rolled steel is stronger and harder than the material from which it is rolled. Nonferrous metals, such as copper and aluminum, also work-harden if they are subjected to rolling, hammering, or cold drawing.

CLASS OF HARDNESS	BRINELL HARDNESS NUMBER	ROCKWELL HARDNESS		SHORE SCLEROSCOPE HARDNESS	APPROX. TENSILE STRENGTH
		C	B		
File	745	67		94	372,000
Hard	682	63		89	341,000
Very Difficult to Machine	627	60		84	317,000
	555	55		75	284,000
	495	50		68	253,000
	444	46		62	221,000
	401	42		56	193,000
	375	39		52	178,000
	352	37		49	165,000
Commercial Machine Range	331	35		46	154,000
	293	30		41	135,000
	255	25		35	117,000
	223	20	97	31	105,000
	179	10	89	25	87,000
	163	6	85	23	81,000
	156	4	83	23	78,000

Fig. 70-4 Hardness conversion table

OTHER CHARACTERISTICS

The melting points shown in figure 70-5 are the points at which metals change to a solid during the cooling cycle. It should be noted that the melting points are listed in Fahrenheit and the balance of the temperature constants in Celsius.

Thermal conductivity, figure 70-5, is a measure of a metal's ability to conduct heat. It seems to have very little relation to the other characteristics except in the matter of *electrical conductivity.* It will be noted from the values in figure 70-5 that those metals having a high-thermal conductivity also have a high-electrical conductivity. Those metals with low-thermal conductivity have a correspondingly low-electrical conductivity.

One of the generally accepted metallurgical rules is that a pure metal is a better conductor than any of its alloys. Silver, which is the accepted standard for thermal and electrical conductivity, is a better conductor of both heat and electricity than an alloy of silver and any other element.

The values in the third column of figure 70-5 are a measure of the metal's resistance to the flow of electricity. The metals with the lower resistance figures are better conductors.

The values for the measurement of *linear expansion* are indicated as millionths-of-an-inch growth or expansion per inch of length per degree Celsius rise in temperature. The value of silver would be expressed as 19.68 x .000001 inch or .00001968 inches per inch

METALS	MELTING POINT OF METALS IN DEGREES F.	THERMAL CONDUCTIVITY CAL./SQ. CM./CM./SEC./ DEGREE C.	ELECTRICAL RESISTIVITY MICROHM/CM.	COEFFICIENT OF LINEAR EXPANSION MICROIN./IN. /DEGREE C.	SPECIFIC HEAT IN CAL./ GRAM/DEGREE C.
SILVER	1761	1.00	1.59	19.68	0.056
COPPER	1981	0.94	1.67	16.5	0.092
ALUMINUM	1220	0.57	2.64	23.6	0.215
LEAD	621	0.08	20.64	29.3	0.03
MAGNESIUM	1202	0.367	4.45	27.1	0.245
NICKEL	2647	0.22	6.84	13.3	0.105
PURE IRON	2797	0.18	9.71	11.76	0.11

Fig. 70-5 Physical constants important to welders Note: Figures for mild steel would be similar to those shown for pure iron.

Linear expansion is always important to the welder. The very nature of the welding process usually causes uneven heating and uneven cooling. This sets up severe thermal stresses in some metals and usually results in rather high *residual stress* when the weldment cools. Residual stress sometimes results in warping and other distortions which make the weld unacceptable.

All users of metals are concerned to some degree with *thermal expansion.* Our standard of measure of length, the meter, is maintained in Washington by the Bureau of Standards. This bar is made of the alloy invar (36 percent nickel and 64 percent iron) which has nearly a zero coefficient of expansion. Makers of high-grade watches and clocks make balance wheels and pendulums of invar to compensate for temperature changes.

Makers of electrodes for arc-welding cast iron frequently make nickel-steel alloy rods of a composition that closely approximates the expansion rate of cast iron. This produces a ductile weld that does not set up excessive stresses when cooled.

Railroad builders leave a gap at the end of each rail section to allow for expansion during the summer heat. Electric power companies construct their high-tension lines with a definite sage between towers. They use charts indicating the among of sag to be provided, depending on the temperature in relation to the distance between towers at the time the lines are run. This sag compensates for the shrinkage that takes place as the atmosphere cools. In some regions a range between a summer temperature of 90 degrees F. (32.2 degrees C) and a winter temperature of –30 degrees F. (–34 degrees C) is not unusual.

Specific heat is a measure of a metal's thermal capacity; the ability to store heat. It is usually expressed as the ratio of the quantity of heat required to raise the temperature of the metal through a given range to the heat required to raise the temperature of an equal mass of water through the same range. It may also be expressed as the amount of heat necessary in calories or Btu's to raise the temperature of a given mass of metal 1 degree Fahrenheit.

WELDING ALUMINUM

Aluminum is considered to be difficult to oxyacetylene weld. Some of the difficulty can be explained by comparing the chart values of aluminum with the values of other metals such as iron or steel; these metals are not considered very difficult to weld in spite of the fact that their melting points are over twice that of aluminum. The problem develops because the thermal conductivity of aluminum is slightly over three times that of iron. Heat applied to the welding zone will be carried away much faster when the material is aluminum.

If the mass of material is large, then the amount of heat required is proportionately large before any melting takes place. The specific heat of aluminum is almost twice that of iron; therefore, almost twice as many calories or Btu's are required to raise the aluminum a given number of degrees.

Another difficulty, not indicated in figure 70-5, is the formation of aluminum oxide during welding. It covers the aluminum with a hard, tough skin that has a melting point in excess of 5,000 degrees F. (2760 degrees C) and requires special fluxes and techniques when torch welding. On the other hand, iron oxide forms more slowly, has a melting point somewhat below that of iron or steel (around 2,000 degrees F./1,093 degrees C), and floats, thus presenting no great problem when torch welding.

EXPERIMENT NO. 1 TESTING WORK HARDNESS

Material

Straight steel gas welding rod, 1/8'' x 36''

Procedure

1. Lay the rod on a flat surface and heat it at the center for about 3 inches to a red heat.
2. Allow the rod to cool in the atmosphere.
3. Apply stress.

Observation and Conclusion

The rod will bend much more easily than before it was heated. This indicates that much of the strength and springiness was imparted to the rod when it was cold drawn.

How does heating affect the hardness of the steel welding rod?

EXPERIMENT NO. 2 TESTING WORK HARDNESS

Materials

Steel plate 2 in. x 8 in. x 1/4 in.
Manganese steel electrode or 18-8 stainless steel electrode
Chisel
Ball peen hammer

Procedure

1. Run a bead on the steel plate.
2. Place the chisel across the weld and strike a hard blow.
3. Peen the bead at some other point until the surface shows evidence of deformation.
4. Repeat step 2 on the peened portion of the bead.

    CAUTION: Wrap cutting edge of chisel in a leather glove as a safety precaution.

Observations and Conclusions

1. Examine and compare the depth of penetration for each cut. Examine the cutting edge of the chisel.
2. What does this experiment indicate about work hardening?
3. What happens if the electrode used is in the softer E-60 series?

EXPERIMENT NO. 3 TESTING HARDNESS

Materials

Cast iron welding rod
Mill file
Permanent magnet
Tank of water

Procedure

1. Heat 3 inches of one end of the welding rod until it becomes bright red (1,650 degrees F./899 degrees C). This transition temperature can be determined by using a permanent magnet.

   Note: When steel and iron reach their upper critical temperature, they are no longer attracted by a magnet. Actually, they lose their magnetic attraction from 50 to 100 degrees F. (10 to 38 degrees C) below their true upper critical temperature.

2. Plunge the hot iron into a tank of water. Keep it moving by stirring until it cools.

3. File the end that was heated with a mill file. Observe how much metal is removed and listen to the sound of the file.

4. Repeat the filing procedure on the other end. Compare the amount of filed metal with that removed from the heated end.

5. Place 1 inch of the hardened end into the jaws of a vise, wrap it with a glove, and bend the rod until it breaks.

6. Break the other end of the rod and examine both fractures.

Observations and Conclusions

1. Compare the amounts of metal filed from the treated and untreated ends of the rod. What are the differences? Why do they exist?

2. What do the two fractures reveal about the chemical compound of iron carbide as compared with the mechanical mixture of ferrite and carbon?

3. If a drill rod were used instead of cast iron, what additional step would be necessary to ensure a fracture in the untreated end with a minimum of time and effort?

EXPERIMENT NO. 4 TESTING THERMAL CONDUCTIVITY

Materials

1 carbon and 1 graphite welding electrode of equal length and diameter.

Procedure

1. Grasp the electrodes in an electrode holder with about 2 inches of each electrode extending below the jaws of the holder.

2. Strike an arc on a piece of scrap and allow each electrode, in turn, to become incandescent (hot enough to glow) along its 2-inch length.

3. After 2 or 3 minutes, check the temperature of the unheated ends of the electrodes. This can be done by touching the presumably cool ends, but the student should use the same caution used if they were hot.

Observations and Conclusions

1. Compare the difference in temperature of two materials to obtain an approximation of the difference in thermal conductivity.

2. It has been said that thermal and electrical conductivity are closely related. Which material is probably the best conductor of electricity? How can one tell the difference between carbon and graphite electrodes?

EXPERIMENT NO. 5 TESTING THERMAL CONDUCTIVITY

Materials

Welding rods of equal length and diameter (1/8 in. diameter x 8 inches long) cut from mild steel, copper, aluminum, brass, and stainless steel

Wax

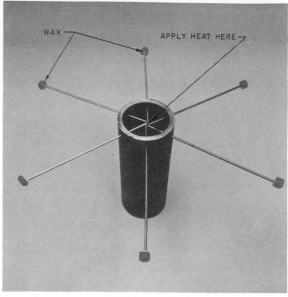

Fig. 70-6 Setup for comparing thermal conductivity

Procedure

1. Attach small pieces of wax to the end of each rod by heating the rod or wax slightly and holding the rod and wax together until the wax adheres to the rod. Insert the blank ends of the rods into the holder until they meet at the center, figure 70-6.

2. Apply heat at the junction of the rods with a welding flame. After a short time, remove the flame and observe the wax.

3. As the heat travels along the rods, the bits of wax will drop off. Record the order in which this occurs. The results will give a comparison of the conductivity of the metals used.

Observation and Conclusion

Which metal tested shows the highest thermal conductivity? Which metal shows the lowest thermal conductivity? How do these results compare with the values given for thermal conductivity in figure 70-5?

EXPERIMENT NO. 6 TESTING SPECIFIC HEAT

Materials

5 small metal samples equal in weight
Thermometer with 200-degree F. range
Beakers and water

Procedure

1. Place the metal samples in a pan of water and boil the water long enough for the metal samples to reach the temperature of the boiling water — about 212 degrees F. (100 degrees C). This will take about 15 minutes.

2. Place the metal samples in separate beakers of water at room temperature — 68 degrees F. (20 degrees C).

3. Using a thermometer with a 200-degree F. range, measure the temperature of the water after it has absorbed enough heat to stabilize metal and water temperatures.

    Note: If several metals are used and a stabilized temperature is recorded for each, you will have a comparison chart of a specific heat of these metals.

Observation and Conclusion

Which metal tested shows the greatest specific heat? Which metal tested shows the lowest specific heat? Compare these values with the values given for thermal conductivity in figure 70-5.

EXPERIMENT NO. 7 TESTING THERMAL EXPANSION AND CONTRACTION

Materials

Steel bar 3/4 in. diameter x 5 in. long
Micrometer

Procedure

1. Measure the diameter and length of the bar with a micrometer. Record measurements.

2. Place the bar in a vise and tighten the vise enough to hold the bar.

3. Heat a portion of the center of the bar to a red heat with a welding flame. Remove the flame and allow the piece to cool. It will eventually drop from the vise.

4. After the bar has returned to room temperature, measure its length and diameter with the micrometer. Compare the new measurements with the original dimensions.

Observations and Conclusions

1. Why did the bar not return to its original length and diameter when it cooled to its original temperature?

2. What would the result be if a bar of comparable dimensions were heated on a welding bench with the material free to expand in all directions?

3. Compare the experiment results and the tendency of metals to warp when welded.

REVIEW QUESTIONS

1. What is the definition of hardness?

2. What is the primary element that governs the hardness of steel?

3. What, if any, advantage is there to using the Rockwell hardness tester over other systems?

4. What is the advantage of using high-manganese steel alloys in cases of high abrasion and impact service?

5. What process makes steel harder and stronger without heat treating?

6. Explain the term thermal conductivity.

7. Of what importance is the knowledge of thermal conductivity to welders?

8. What characteristics, in addition to thermal conductivity, makes a metal difficult to bring to the melting point?

9. Lead melts at 621 degrees F. (327 degrees C), yet many welders experience much difficulty in trying to weld it. Why?

10. Stainless steel has about the same coefficient of linear expansion as copper. Stainless steel has a much lower index of thermal conductivity than copper. Is stainless steel a good or bad material to use for backing bars, and jigs and fixtures when clamping and holding copper during welding operations? Why or why not?

11. If the welding rods used in the experiment using the setup in figure 70-6 were magnesium, aluminum, copper, lead, iron, and nickel, in what order would the wax drop from the rods after heating as directed?

UNIT 71 THE EFFECTS OF HEATING AND COOLING CYCLES ON METALS

ALLOY STEELS

Metal fabricating, including welding, involves the heating and cooling of alloys, particularly those composed primarily of steel. All steels contain iron, carbon, manganese, silicon, sulphur, and phosphorus and, technically, would be termed alloys by the definition. *Plain carbon steels* (no alloy added) are not generally thought of as an alloy.

The alloy steels are roughly classified into *low-alloy steel* and *alloy steel.* The low-alloy steels are an intermediate group between carbon steel and alloy steel; low-alloy steels have a small percentage of alloying elements such as manganese, nickel, molybdenum, and chrome. They have been developed to provide a steel of intermediate price with a high strength-to-weight ratio. They are used in the transporation industry where these alloys provide a greater weight-carrying capacity with no increase in the weight of the carrier.

If strength and hardness were the only considerations, plain carbon steel would produce very high-strength steel of great hardness. However, alloying elements, such as those indicated in the SAE numbering systems, are added to steel to secure other beneficial characteristics, as displayed in figure 71-1.

CHARACTERISTICS	APPLICATIONS
Combinations of strength and hardness	Shovel teeth, rock and crusher jaws
Ductility	For deep drawing of material that makes up auto body parts; the seamless drawn cylinders that contain oxygen at high pressures
Deeper penetration of heat-treating	Heat-treating for high-speed tool bits and dies
Retention of hardness at high temperatures	For high-speed tools using tungsten carbide or nonferrous alloys which retain their cutting edge even near a red heat
Resistance to corrosion or scaling at high temperatures	Equipment used to produce and heat-treat steel; in chemical and metal laboratories; and exhaust rings and after-burners in jet engines
High resistance to flow of electricity	For heating elements in electric furnaces, flat irons, electric stoves, and toasters
Magnetic permeability	For electric motors, transformers, and electro-magnets used in lifting magnets and holding magnets

Fig. 71-1

HEAT TREATING

Heat treating is defined as heating and cooling finished metals or alloys to produce certain desirable properties such as strength, hardness, softness, ductility, and grain refinement. Heat treating can be summarized by four statements:

- Steel undergoes definite internal changes when subjected to temperatures above its critical range.

- After reaching this temperature, if the steel is allowed to cool naturally, it will tend to return to its normal condition similar to that found after normalizing.

- For steel to return to normal condition, sufficient time must elapse during the cooling cycle so that the internal changes that took place during heating have time to reverse themselves.

- If heated steel is cooled at a rate which the internal changes can reverse themselves, this will produce certain modifications in the structure of the metal. These structural changes alter its physical characteristics, such as strength, hardness, toughness, and ductility.

TEMPERATURE INDICATORS

Convenient temperature-indicating crayons, with the melting point marked in degrees, are frequently used in welding shops. For instance, if a mark is made on a piece of metal with a crayon labelled 700 degrees F. (371 degrees C) and the piece is heated, the mark will melt when the temperature reaches 700 degrees F. (371 degrees C). This approximates the annealing (softening of metals by using heat) temperature for aluminum. If the crayon is rated at 1,500 degrees F. (816 degrees C), the mark melts at approximately the upper critical point of carbon steel.

The *critical range* and *critical points* are designated as Ac_1, Ac_2, and Ac_3 for heating and Ar_3, Ar_2, and Ar_1 for cooling. These letters were taken from the French language.

$$A = Arret \text{ (stop)}$$
$$C = Chauffage \text{ (heating)}$$
$$R = Refroidissment \text{ (cooling)}$$

Thus, Ac_1 = Stop heating at the number 1 or critical point. In the case of falling temperatures, these points are referred to in the order of their occurrence as Ar_3, Ar_2, or Ar_1. Thus, Ar_1 means "stop cooling at a lower critical point."

CRITICAL RANGE

Heating operations usually involve the use of temperatures in the critical range and slightly above or below this range. Relatively low temperatures are sometimes used in tempering and drawing. If a piece of SAE-1030 steel is heated, its color will change through the temper colors and up into the red range, becoming brighter as the temperature increases. At the Ac_1 point, 1,350 degrees F. (732 degrees C), the color will remain constant for a short time even though heat is being applied. Up to this point the metal will expand at a uniform rate, proportionate to the temperature.

Basic Guide to Ferrous Metallurgy

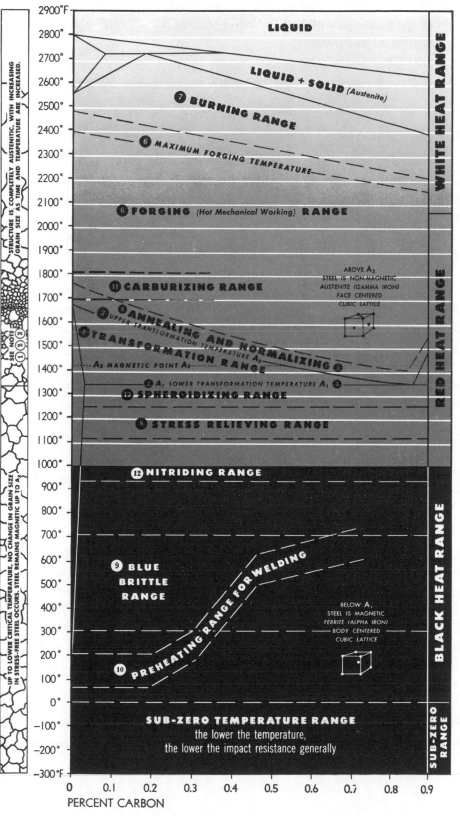

① TRANSFORMATION RANGE. In this range steels undergo internal atomic changes which radically affect the properties of the material.

② LOWER TRANSFORMATION TEMPERATURE (A_1). Termed Ac_1 on heating, Ar_1 on cooling. Below Ac_1 structure ordinarily consists of FERRITE and PEARLITE (see below). On heating through Ac_1 these constituents begin to dissolve in each other to form AUSTENITE (see below) which is non-magnetic. This dissolving action continues on heating through the TRANSFORMATION RANGE until the solid solution is complete at the upper transformation temperature.

③ UPPER TRANSFORMATION TEMPERATURE (A_3). Termed Ac_3 on heating, Ar_3 on cooling. Above this temperature the structure consists wholly of AUSTENITE which coarsens with increasing time and temperature. Upper transformation temperature is lowered as carbon increases to 0.85% (eutectoid point).

● FERRITE is practically pure iron (in plain carbon steels) existing below the lower transformation temperature. It is magnetic and has very slight solid solubility for carbon.

● PEARLITE is a mechanical mixture of FERRITE and CEMENTITE.

● CEMENTITE or IRON CARBIDE is a compound of iron and carbon, Fe_3C.

● AUSTENITE is the non-magnetic form of iron and has the power to dissolve carbon and alloying elements.

④ ANNEALING, frequently referred to as FULL ANNEALING, consists of heating steels to slightly above Ac_3, holding for AUSTENITE to form, then *slowly* cooling in order to produce small grain size, softness, good ductility and other desirable properties. On cooling slowly the AUSTENITE transforms to FERRITE and PEARLITE.

⑤ NORMALIZING consists of heating steels to slightly above Ac_3, holding for AUSTENITE to form, then followed by cooling (in still air). On cooling, AUSTENITE transforms giving somewhat higher strength and hardness and slightly less ductility than in annealing.

⑥ FORGING RANGE extends to several hundred degrees above the UPPER TRANSFORMATION TEMPERATURE.

⑦ BURNING RANGE is above the FORGING RANGE. Burned steel is ruined and *cannot be cured* except by remelting.

⑧ STRESS RELIEVING consists of heating to a point below the LOWER TRANSFORMATION TEMPERATURE, A_1, holding for a sufficiently long period to relieve locked-up stresses, then slowly cooling. This process is sometimes called PROCESS ANNEALING.

⑨ BLUE BRITTLE RANGE occurs approximately from 300° to 700°F. Peening or working of steels should not be done between these temperatures, since they are more brittle in this range than above or below it.

⑩ PREHEATING FOR WELDING is carried out to prevent crack formation. See TEMPIL° PREHEATING CHART for recommended temperature for various steels and non-ferrous metals.

⑪ CARBURIZING consists of dissolving carbon into surface of steel by heating to above transformation range in presence of carburizing compounds.

⑫ NITRIDING consists of heating certain *special steels* to about 1000°F for long periods in the presence of ammonia gas. Nitrogen is absorbed into the surface to produce extremely hard "skins".

⑬ SPHEROIDIZING consists of heating to just below the lower transformation temperature, A_1, for a sufficient length of time to put the CEMENTITE constituent of PEARLITE into globular form. This produces softness and in many cases good machinability.

● MARTENSITE is the hardest of the transformation products of AUSTENITE and is formed only on cooling below a certain temperature known as the M_s temperature (about 400° to 600°F for carbon steels). Cooling to this temperature must be sufficiently rapid to prevent AUSTENITE from transforming to softer constituents at higher temperatures.

● EUTECTOID STEEL contains approximately 0.85% carbon.

● FLAKING occurs in many alloy steels and is a defect characterized by localized micro-cracking and "flake-like" fracturing. It is usually attributed to hydrogen bursts. Cure consists of cycle cooling to at least 600°F before air-cooling.

● OPEN OR RIMMING STEEL has not been completely deoxidized and the ingot solidifies with a sound surface ("rim") and a core portion containing blowholes which are welded in subsequent hot rolling.

● KILLED STEEL has been deoxidized at least sufficiently to solidify without appreciable gas evolution.

● SEMI-KILLED STEEL has been partially deoxidized to reduce solidification shrinkage in the ingot.

● A SIMPLE RULE: Brinell Hardness divided by two, times 1000, equals approximate Tensile Strength in pounds per square inch. (200 Brinell $\div$ 2 $\times$ 1000 $=$ approx. 100,000 Tensile Strength, p.s.i.)

Copyright 1954, **TEMPIL** CORPORATION.
132 West 22nd Street • New York 11, N. Y.

At the Ac_1 point, the expansion will halt and the metal will begin to shrink, and will continue to shrink until the Ac_3 point is reached (1,495 degrees F./813 degrees C) after which it will return to its normal expansion rate. When steel reaches or goes beyond its upper critical temperature it becomes nonmagnetic.

Note: These critical points will vary somewhat with the carbon content. Steels having a higher carbon content, in general, will have lower critical points up to about .85 percent carbon.

STRUCTURAL AND CHEMICAL CHANGES

Certain structural and chemical changes take place in the heating and cooling cycles. If the 1030 steel bar is examined under a microscope, it will be found to be composed almost entirely of ferrite and pearlite. The pearlite is composed of alternate layers of ferrite and cementite. The cementite is one of the iron carbides — a hard chemical combination of iron and carbon. When this bar is heated, no change will take place until the metal reaches the Ac_1 point. At this temperature, the ferrite begins to act as a solvent in which all the carbide goes into solution. All the steel still remains solid, and the resulting chemical combination is called a solid solution. This change continues and is completed as the steel reaches the Ac_3 point when all the carbon is in solution with the iron. This combination is called *austenite.* Austenitic steel is nonmagnetic. Figure 71-2 shows several of the important temperatures for ferrous metals.

It is important to note that these changes do not take place immediately. A certain amount of time is necessary for the transformation to take place. It is by taking advantage of this time lag that we are able to harden steel.

When steel is cooled rapidly (quenched) from above the Ac_3 point, the austenite does not have time to return to ferrite and pearlite, but changes instead into a substance known as *martensite,* the hardest and most brittle of the iron — iron carbide structures.

MICROSCOPIC EXAMINATION

Microscopic and submicroscopic studies of metals which have been heated and cooled at different temperatures help to understand the changes that take place. The sample is prepared for examination by polishing it to a mirror finish and etching it with the proper reagent. A 5-percent mixture of nitric acid in alcohol, nital, is a common etching solution for steel. The etched section is placed under a metallurgical microscope with resolving power of 100 to 2,500. The individual crystals of the parts can be observed and their size, arrangement, and structure noted.

Metallurgical laboratories use metallographs to magnify and photograph microstructures. Electron microscopes may be used to magnify the grains up to 25,000 x magnification. X-rays are also used to investigate the atomic structure of the metals.

EXPERIMENT NO. 1 TESTING EFFECTS OF HEATING ON GRAIN STRUCTURE

Materials

3/16-in. steel plate (or thicker)
Welding tips

Procedure

1. Make two fillet welds on the steel plate: one by the oxyacetylene process and the other by the electric-arc process and allow welds to cool. Which weld cooled faster?

2. Break each weld by hammering the upstanding leg of the joint toward the weld.

Observation and Conclusion

Compare the grain size shown in the fracture. What does this tell you about the effect of the rate of cooling on grain size?

EXPERIMENT NO. 2 TESTING EFFECTS OF HEATING ON GRAIN STRUCTURE

Materials

3/4 in. to 1 in. thick steel plate
Alcohol
Ammonium persulphate
Clear lacquer

Procedure

1. Make a beveled butt joint by arc-welding the steel plates. Use the multiple-pass technique with stringer beads or relatively thin weave beads.

2. Make another butt weld using larger electrodes, higher current value, and a slower rate of travel so that the joint may be completed in one or two passes. Make the weld with two beads if possible.

3. Cross section each weld and polish each coupon by first grinding and then stroking the coupon on succeeding finer grades of abrasive cloth placed on a flat surface until it takes on a mirror finish.

4. Etch coupons with a reagent such as a saturated solution of ammonium persulphate.
   Note: This type of etching is known as *macroetching* and differs from microetching in that results can be observed with the naked eye (figure 71-3). Microetching requires a microscope for observation.

5. To prevent the coupons from oxidizing, wash them in hot water immediately after etching and then place them in alcohol to remove the water. For long-term preservation, coat the clean, dry surface with clear lacquer.
   CAUTION: Ammonium persulphate is one of the safest etching chemicals, but it should be kept away from the eyes and mouth at all times. If any solution touches the skin, the area should be immediately flushed with water.

Observations and Conclusions

1. Observe each etched coupon and compare the structure of the finished welds. Examine and compare the grain structure of the underlying beads with each other and with the final bead. Is there any similarity between the etched coupons and that shown in figure 71-3?

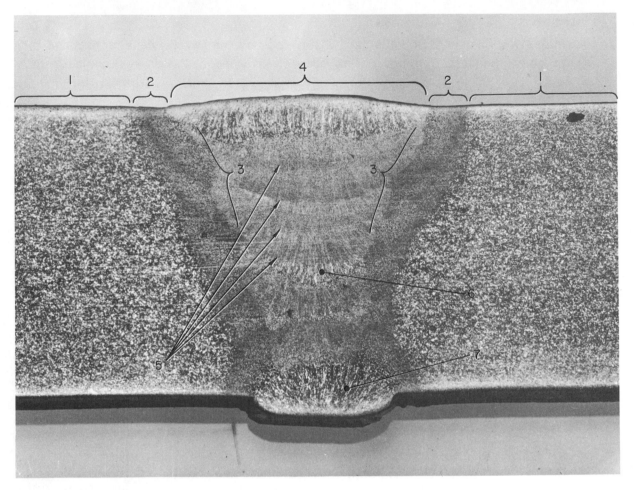

1. Grain structure of parent metal — 1025 steel

2. Heat-affected zone. Note finer grain structure.

3. Weld boundary

4. Final pass of weave welds. Note coarse, columnar grain structure.

5. Successive bead boundaries. Note that each successive bead refines the grain structure of the underlying beads.

6. If the pass is thick enough, heat from succeeding beads does not penetrate deep enough to refine grain structure of entire bead, leaving some coarse grain structure.

7. Final bead. Note coarse, columnar grain with grain growing radially from colder metal toward the last part of bead to "freeze out," or solidify.

Fig. 71-3 Macroetched section of multiple-pass weld

2. When steel cools rather slowly and recrystallizes from the molten state, each grain forms and attaches itself to a previously formed grain in a pattern resembling a pine tree — hence, the term *dendrite* (branchlike). Why is there no dendritic structure in the underlying beads in the multiple-pass weld?

3. If the weld made with the slower rate of travel has only two beads, what is observed in the underlying bead? Is there any evidence of dendritic structure in the bead? If so, why does this condition exist?

REVIEW QUESTIONS

1. Other than temperature, what is the most important factor when heat-treating steel and its alloys?

2. The text indicates that steel expands at a uniform rate to the Ac_1 point and then shrinks until the Ac_3 point is reached. After the Ac_3 point, it returns to its normal rate of expansion. How is the phenomenon explained?

3. The term austenitic steel is often found in texts, handbooks, and advertisements. What does this term mean?

4. When the welds made by the oxyacetylene process are broken what visible difference is there in the grain size? What causes this?

5. After the cross section of a multiple-pass weld has been macroetched, how does the final bead look? The underlying beads? What causes this change?

6. When steels are hardened, they are quenched, or cooled rapidly from their critical range. How does this effect the steel?

UNIT 72 CHANGING PROPERTIES OF STEEL

Besides the heating and cooling process for changing the characteristics of steel and its alloys, the welder encounters other heating and cooling effects. These effects occur in a variety of gaseous and solid atmospheres and alter the welding procedures that must be followed. An elementary knowledge of nitriding, carburizing, malleableising, stress-relieving, normalizing, annealing, and some of the causes and effects of grain growth of metals is needed.

Steel and iron at high temperatures exhibit at least two important characteristics which in some cases, such as malleableizing, are advantageous and in other cases, such as annealing may be harmful.

First, steel and iron at elevated temperatures tend to absorb many of the elements that make up the surrounding atmosphere.

Secondly, carbon tends to migrate, even in the solid solution, from areas of high-carbon content to areas of low-carbon content. The rate and degree of migration is in direct proportion to the temperature of the material and the length of time the temperature is maintained.

MALLEABLEIZING

In the case of malleableizing, the basic material is white cast iron, which exhibits a white fracture because all of the carbon is in solution as iron carbide. Gray cast iron exhibits a gray fracture because some of the carbon is separated from the iron in the form of carbon flakes interspersed among the grains of iron and iron carbide.

White cast iron may be malleableized by heating the iron above the Ac_3 point. Some of the carbon at and near the surface migrates and combines with the oxygen in the furnace atmosphere.

When malleable cast iron is broken, the fracture shows a bright, steely appearance from the surface inward. The rest of the casting is gray in appearance because the carbon has separated from the ferrite to form temper carbon. This leaves the entire casting in a soft, somewhat ductile condition. It can be bent or hammered to some extent without fracturing.

The other method of malleableizing white cast iron is to pack the casting in a box with the entire piece surrounded by *mill scale,* which is iron oxide. When the iron is heated at or above the Ac_3 point, the carbon migrates to the surface and combines with the oxygen in the iron oxide. The depth to which this reaction takes place is dependent on the length of time the material is held at the indicated temperature. The resulting product can be described as a semisteel with considerably more resistance to bending and shock.

If a welder were to attempt to repair a broken malleable casting by welding, the original heat-treating temperature would be exceeded and all the elements present would recombine at or above the melting point. The final product would be a grade of gray cast iron in and near the welding zone. For this reason, malleable castings are always brazed to keep the temperature low enough to avoid chemical or mechanical changes.

Decarburization (the loss of carbon from the surface of steel), which is an advantage when producing malleable iron, is a definite disadvantage when trying to harden steel by heat-treating. The carbon in the steel combines with the oxygen in the furnace atmosphere and leaves the surface of the finished product with a skin of lower carbon, softer steel.

CARBURIZING OR CASE-HARDENING

In many instances steel needs to have a hard wear-resisting surface and a ductile shock-resistant core. An example is the inner cones and outer races of wheel bearings in automobiles. A soft bearing would not wear well and a totally hard bearing would shatter. The wear-resistant surface is produced by *case-hardening* (surface-hardening) in the carburizing or nitriding bath process.

The carburizing process is somewhat the reverse of the malleableizing process. The low-carbon steel is heated in a carbon-rich atmosphere, either in boxes packed with solid material containing carbon, or in rotating heat-resisting cylinders filled with a carbon-rich gas. The exact temperature depends on the process used.

In the *pack-hardening* process, the temperature is usually slightly above the Ac_3 point. In the rotating-cylinder method, the temperature is usually kept lower to eliminate the tendency of the parts to distort as they are tumbled. In either case the steel absorbs carbon up to as much as 1.2 percent at the surface, and the pieces are quenched directly from the furnace to keep them bright. The depth of this hard case varies with the length of time the work is exposed to the hot, carbon-rich atmosphere.

The student can check the effects of carburizing by grinding one face of a discarded bearing race and placing it in a beaker containing hydrochloric acid. After a few hours it can be seen that the acid has etched the soft core more than the hard exterior.

Other methods of carburizing and nitriding consist of heating the work in molten baths of sodium cyanide mixtures or other salts containing either carbon or nitrogen. In this process the depth of the case is usually less than that produced by other case-hardening processes. A case of 0.01 inch is general, and one as high as 0.03 is rare. Cyaniding and other salt bath case-hardening processes can be extremely dangerous since cyanide is a deadly poison. The principal element producing a hard case in cyaniding is carbon.

The nitriding process accomplishes the case-hardening by adding nitrogen to the surface of the steel in a molten bath of nitrogen-rich salts and, in some instances, by heating the steel in a furnace atmosphere of ammonia.

Welders often take advantage of steel's ability to absorb carbon by heating a piece that is to be hard-faced with a carburizing flame. This adds carbon to the surface and lowers the melting point. A thin skin of the work is melted and, as the expensive hard-facing material is applied, there is very little mixing of the base metal and alloy. Therefore, the content of the overlay is not greatly changed.

STRESS RELIEVING BY HEAT

Stress relieving is a process that is commonly encountered in the welding field. The uneven heating and cooling of the welding process sets up stresses in the work. Those remaining when the work has cooled are called *residual stresses.* These stresses sometimes cause no problem and can be ignored. In other cases they cause unwanted or harmful distortion and must be eliminated.

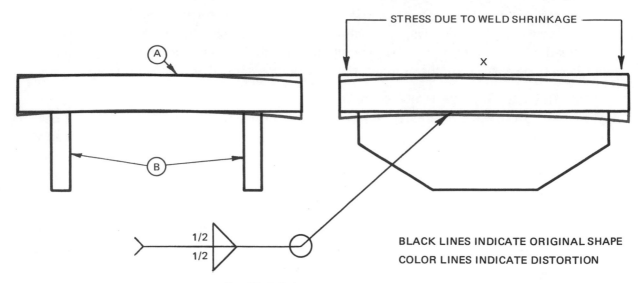

X

BLACK LINES INDICATE ORIGINAL SHAPE

COLOR LINES INDICATE DISTORTION

1/2

1/2

Fig. 72-1 Effect of residual stresses

In figure 72-1 all welding has been done on one side of the weldment, and lines in color indicate the shape the work would tend to assume. If the proper amount of restraint had been applied during the welding and cooling cycle, the surface of plate A would present a perfectly flat surface. If plate A and ribs B had been predeformed to compensate for the weld shrinkage, then the residual stresses would pull the weldment into the desired shape. If the weldment is to be used in the welded condition there is no problem. However, if the top of plate A must be machined, the top surface becomes weaker as the metal is removed and the residual stresses tend to distort or bow the surface. The more metal that is removed, the greater the distortion. Stress relieving is needed.

A weldment is stress relieved by heating it in a furnace to a temperature of not less than 1,100 degrees F. (593 degrees C), and not more than slightly below the Ac_1 point. The generally accepted stress-relieving temperature is 1,200 degrees F. (649 degrees C). The work is held at this temperature for 1 hour for each inch of thickness of the thickest piece of the weldment. Then it is allowed to cool in the furnace until it reaches 600 degrees F. (316 degrees C) or below. The temperatures specified are low enough so that no metallurgical changes take place, but they are high enough for the work to become somewhat plastic, thus allowing the crystals to rearrange themselves into a stress-free alignment.

STRESS RELIEVING BY MECHANICAL METHODS

In many instances heat is not necessary to accomplish simple and adequate stress relieving. Hammering a weld may be all that is necessary. A common hand hammer may be used, but an air hammer fitted with a suitably shaped tool is faster and allows greater flexibility. Whatever the mechanical method used, the weld will be stretched, and some of the stresses set up by the cooling and shrinking at the time of welding will be relieved. Naturally, this cold working will increase the hardness and tensile strength to some degree. Mechanical stress relieving does not work well on metals and alloys that have a low ductility and are too hard for the crystals to rearrange themselves under hammer blows.

NORMALIZING

Normalizing, as the name implies, is a process to return hot-rolled steel to its normal condition (the condition it was in when it left the rolling mill and was allowed to cool in still air). Subsequent manufacturing processes applied to the steel affect it, and in order to return it to its normal condition the steel must be heated to a temperature considerably above the Ac_3 point and then allowed to cool in still air.

The object of normalizing is to improve the grain structure. The fairly rapid air cooling leaves the metal slightly harder and higher in tensile strength than metals treated by other processes. Normalized steel is generally free from residual stresses.

ANNEALING

Annealing is a process in which the metal is heated to a temperature of 35 to 100 degrees F. (2 to 38 degrees C) above the Ac_3 point and then cooled very slowly. Steel is annealed for one or both of the following reasons:

- To refine the structure of the metal that may have been altered by previous heat treating.
- To soften the steel so that it may be more easily machined or cold-formed.

Stress relieving, normalizing, and annealing are terms that refer to different processes. None of the three should be loosely used to described either of the other two processes.

SPHEROIDIZING

Spheroidizing is a type of annealing in which the holding temperature is either slightly above or below the lower critical range. When this temperature has been maintained for a sufficient time and the work cooled very slowly, the structure of the metal is not *lamellar* (platelike). The cementite assumes a globular, or *spheroid,* shape which improves the machinability of the metal. In the case of some of the high-speed, air-hardening alloys, the annealing cycle is measured in hundreds of hours and undoubtedly accounts for much of the high cost of these metals and alloys.

TEMPERING AND COLORS

While hardness is a desirable quality, brittleness is not. The quenched steel must be *tempered* or drawn by reheating to change the martensite structure and make the steel softer and more ductile. The hardness and brittleness of steel can be modified only by rc heating it to an intermediate temperature (below the Ac_1 point). The exact temperature is determined by the physical characteristics desired in the finished product.

Fig. 72-2 Temper color chart

Much of the tempering or drawing of carbon steel takes place in the 400 to 600 degrees F. (204 to 316 degrees C) range. As the temperature is increased, the steel becomes somewhat less hard but more ductile and more resistant to impact.

Most production tempering is carried out in furnaces in which temperatures can be accurately controlled. The oldtime blacksmith and even some of the modern-day heat-treaters determine the temperature of a piece of steel by the temper colors. As polished steel is heated, it begins to oxidize at about 410 degrees F. (210 degrees C). The first color of the very thin oxide film is a very light yellow or straw color. As the temperature increases, the thickness of the oxide film increases and the color becomes darker through light straw, dark straw, straw and purple spots, light purple, and dark purple to a final blue-gray color.

The temper color chart in figure 72-2 was made by students. Coupons (sample pieces) of polished 1/8-inch thick steel were sheared and stamped with numbers from 400 to 600 degrees F. in 10-degree F. increments. Then they were placed in an electric furnace at the thermocouple level and arranged so that they could be removed in order, such as 400 degrees F., 410 degrees F., and 420 degrees F.

The furnace was set for 400 degrees F., and all of the pieces were allowed to soak at this temperature. Then the 400-degree F. coupon was removed and allowed to cool while the furnace controls were set at 410 degrees F. and the pieces soaked at this temperature for about 10 minutes. The 410-degree F. coupon was removed, and the process repeated at 420 degrees F., 430 degrees F., and so forth.

The pieces were then drilled and attached to a polished board to provide an accurate, attractive temper color chart.

EXPERIMENT: STRESS RELIEVING BY MECHANICAL METHOD

Material

Sheet of steel approximately 2 in. wide x 12 in. long x 1/16 in. thick

Procedure

1. Heat one edge of the steel sheet with an oxyacetylene flame. Move the flame rapidly so that the edge becomes red hot for only about 1/2 in. width.
2. Allow the piece to cool. Observe distortion caused by uneven heating and cooling.
3. Using a hammer and anvil, peen the pieces repeatedly, gently, and evenly, along the heat-affected edge. Try to relieve the thermal stresses just enough to return the work to its original straight, flat condition.

Observation and Conclusion

What happens if you peen the work too hard or too much? What happens if you peen the work more at one spot than at others?

REVIEW QUESTIONS

1. Define malleableizing.

2. Since carbon migrates from areas of high carbon to areas of low carbon even in solid solution, would it be possible to weld one piece of high carbon and one piece of low-carbon steel and heat them so that the carbon content became about equal throughout the entire mass?

3. If a high-carbon and a low-carbon material need to be joined by arc welding, would an electrode of high or low carbon be used? Why?

4. Why is malleable iron always brazed?

5. How deep is the case or hardened zone when steel is carburized or case hardened?

6. What is the purpose in using a reducing or carburizing flame when applying a hard overlay to steel? What is a secondary benefit?

7. Describe thermal stress relieving.

8. How does stress relieving differ from annealing?

9. Is mechanical or thermal stress relieving more effective? Why?

10. If some stress-relieved weldments were mixed with identical weldments which had not been stress relieved, how could they be sorted?

11. Specifications often call for welded pressure vessels to be normalized before they are put in to use. What would happen if a riveted pressure vessel were subjected to the same treatment?

12. Why does a welder need to be able to recognize the temper colors?

UNIT 73 METHODS OF IDENTIFYING METALS

THE PROBLEM OF RECOGNITION

The process of identifying metals ranges from simple to complex. Some persons with practically no knowledge of metals may have no trouble distinguishing between such metals and alloys as copper, brass, and aluminum simply by observing their color. On the other hand, if pieces of steel or iron were electroplated with copper or brass, color would not indicate the underlying metal, nor would picking up the pieces to check their weight aid greatly. One would feel about as heavy as the other. In this case a magnet would distinguish between copper and brass, which are nonmagnetic, and the coated steel, which would be attracted by the magnet. If other metals such as zinc, white metal, and lead were electroplated, identification would be more difficult because they are nonmagnetic metals and alloys.

Cast iron and plain carbon steels can be identified by observing such simple evidence as rust on the surface. This indicates a ferrous metal or alloy. Even in this case, the absence of rust does not indicate that the alloy contains no iron. The surface may be coated to prevent rust, or the alloy may be designed to be rustproof as in the case of stainless steel.

LABORATORY METHODS OF IDENTIFYING METALS

Chemical Analysis

An exact chemical analysis of a metal or alloy requires the services of a highly skilled chemist or technician using a variety of chemicals and techniques that may call for the use of expensive and sophisticated equipment. Chemical analysis usually takes much time and always requires a high degree of attention to detail at each of the many steps in the procedure.

Optical Devices

A much more rapid means of analysis is the *spectroscope*, which is a high-precision optical device. Its operation is based on the fact that when any of the elements are brought to incandescence, usually by the use of an electric arc, each element emits light rays of a different wave length. When these light rays are passed through a precisely arranged lens and diffraction grating (a device that separates light rays) and projected onto a photographic film, a permanent record is made such as in figure 73-1.

The series of vertical dark lines are termed reinforcing bands and vary in width and location on the film, depending on the element emitting the rays.

Specific Gravity

In many instances, a metal alloy may be identified by finding the specific gravity of the unknown material and checking this against a chart of the specific gravities of the elements. The procedure requires the use of a precision-made balance scale if the results are to be dependable.

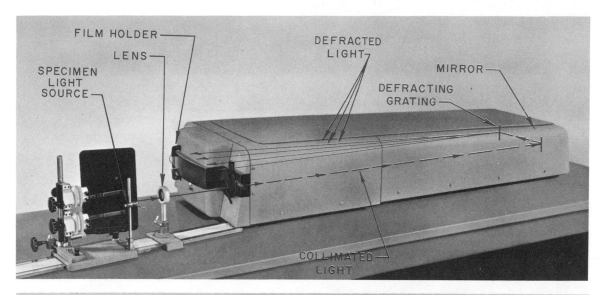

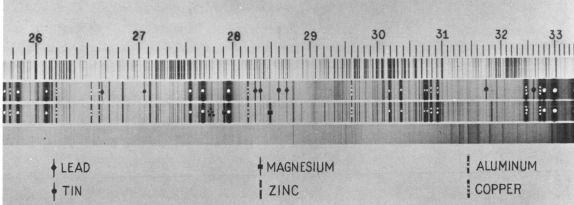

Fig. 73-1 One and one-half meter spectroscope (top). Spectrograph of two die casting alloys showing arrangement of reinforcing bands (bottom).

SHOP METHODS OF IDENTIFYING METALS

Welders and many others engaged in metal fabricating normally do not have access to chemical or physical laboratories and must devise other means for testing and identifying metals. Any one or all of the five senses may be involved in the identification.

Shop methods include identification by physical properties, by examination of fractures and welding tests, by chip testing, by color testing, and by spark testing.

Identifying Lead & Zinc by Physical Properties

If zinc or lead are heated until they oxidize, each will give off a definite and different odor. If the fumes reach the mouth, they taste different. If the materials are heated to the point of rapid oxidation, or burning, the color of the flame is different and the oxide residue is of different colors. The actual colors of the metals themselves are slightly different. If the zinc or lead is not too thick it can be bent, and it can be found that the lead is softer and bends more easily than zinc.

Examination of Fractures and Welding Tests

In general, the welder has available metals which have been fractured or broken. Observe the fracture for grain structure and color; the behavior of the metal under the heat of the torch or arc; and the ease or difficulty of the melting and the appearance of the molten metal. At a lower temperature, surface oxides for color and the extent of the heat-affected zone when a torch is directed at a corner of the material.

The Chip Test

Sometimes a simple chip test with a hammer and chisel is sufficient. For example, cast iron and cast steel have a very similar surface appearance. The cast iron will chip from the work in small pieces rather readily. The steel under similar treatment produces relatively long chips that tend to curl and eventually tear from the work, especially if the test piece is low in carbon content.

The Color Test

A visitor to a welding shop would have no trouble noting the differences in the color of the light reflected from the walls and ceiling of an arc-welding booth. If steel is being welded, the light is predominately blue. If the work is aluminum, there is a pink cast. Copper under the arc produces a green color.

Welders engaged in the TIG welding of aluminum and magnesium and their alloys learn to distinguish those alloys that contain zinc by the purple appearance of the light near the electrode. The size and intensity of this purple light give the welder an indication of the zinc content and whether or not the workpieces can be successfully joined by the welding process. Zinc aluminum and zinc magnesium alloys are not generally recommended for welding.

Spark Test

Probably the most widely used method of identifying metals is the spark test. Many branches of the metals industry employ specialists who are highly skilled in identifying metals by the appearance of the spark given off when an unknown metal or alloy is ground on a grinding wheel. Steel producers who use scrap metal almost exclusively and who produce special alloys of highly accurate analysis, employ skilled spark testers to check and classify all the metals that go into each furnace charge. Other users of metals employ spark testers to check each new shipment of material rapidly to be sure that it meets their specifications. They also make laboratory tests on material selected at random from each shipment, but they rely on the experience and skill of their spark testers.

The key word in spark testing is experience — in fact, the key word in metals identification is experience. No amount of description, diagrams, or pictures can substitute for the actual performance of tests, nor can the performance of tests and observations on a one-time basis develop the skill, knowledge, and techniques necessary to identify metals with any degree of accuracy.

Fortunately, welding students generally have a grinding wheel for spark testing along with a variety of different metals in the form of welding rods which are already identified.

They also have available a number of different metals that are to be used for practice welding. They usually have broken and worn castings of various alloys. By examining surface texture and color, fracture color, and grain size as well as reaction under heating and melting, students can build up a considerable store of knowledge helpful in metal identification.

Such things as broken, high-speed, power hacksaw blades are discarded as useless, but they can be spark tested to gain experience in identifying at least that grade of high-speed steel. Worn-out files make excellent test pieces for observing the spark stream of expectionally high-carbon steel. Hammer-struck tools such as cold chisels tend to mushroom on the struck end and must be ground to make them safe. If the spark stream is observed and compared during these grinding operations, experience will be gained in identifying these alloys.

Many things can be noticed when observing a spark stream. The *color* ranges from dark red for some metals through a variety of shadings of red, orange, yellow and, in a few metals such as titanium, a brilliant white. The color within the spark stream varies with some metals, but it is usually darker near the wheel.

The *length* of the spark stream varies with the material from very short to very long, although some of this variation is due to differences in pressure between the work and wheel. The *total area* of the spark stream varies with the kind of material used, as well as the amount of material in contact with the wheel. The sparks themselves assume several distinct shapes and are described by such terms as appendages, bud break, arrow, dashes, forks, shafts, sprigs, and stars, figure 73-2.

Considering the great variety of combinations possible in the types of sparks and their color and length, it is easy to understand how so many metals and alloys can be identified with a reasonable degree of accuracy, figure 73-3. This is also why it takes considerable time, concentration, and practice to become even semiskilled in spark-testing identification.

Some time can be profitably spent learning to tell the difference between the plain carbon steels by obtaining some low-carbon steel (plate, bar, or structural), some medium-carbon steel (such as 40 percent C-machinery steel), and some high-carbon steel (such as

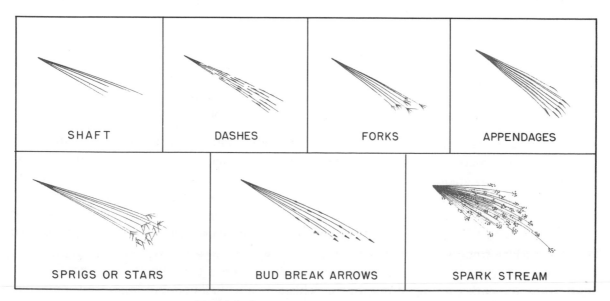

SHAFT   DASHES   FORKS   APPENDAGES

SPRIGS OR STARS   BUD BREAK ARROWS   SPARK STREAM

Fig. 73-2 Components of the spark stream

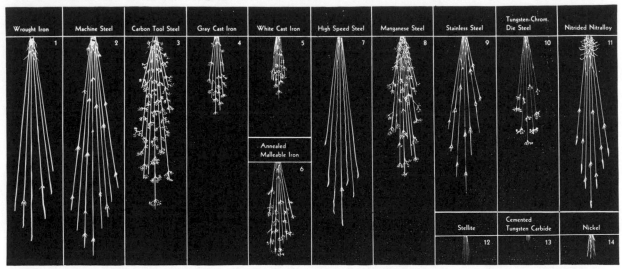

ABRASIVES **NORTON** GRINDING WHEELS

Characteristics of Sparks Generated by the Grinding of Metals

Metal	Volume of Stream	Relative Length of Stream, Inches†	Color of Stream Close to Wheel	Color of Streaks Near End of Stream	Quantity of Spurts	Nature of Spurts
1. Wrought iron	Large	65	Straw	White	Very few	Forked
2. Machine steel (AISI 1020)	Large	70	White	White	Few	Forked
3. Carbon tool steel	Moderately large	55	White	White	Very many	Fine, repeating
4. Gray cast iron	Small	25	Red	Straw	Many	Fine, repeating
5. White cast iron	Very small	20	Red	Straw	Few	Fine, repeating
6. Annealed mall. iron	Moderate	30	Red	Straw	Many	Fine, repeating
7. High speed steel (18-4-1)	Small	60	Red	Straw	Extremely few	Forked
8. Austenitic manganese steel	Moderately large	45	White	White	Many	Fine, repeating
9. Stainless steel (Type 410)	Moderate	50	Straw	White	Moderate	Forked
10. Tungsten-chromium die steel	Small	35	Red	Straw*	Many	Fine, repeating*
11. Nitrided Nitralloy	Large (curved)	55	White	White	Moderate	Forked
12. Stellite	Very small	10	Orange	Orange	None	
13. Cemented tungsten carbide	Extremely small	2	Light Orange	Light Orange	None	
14. Nickel	Very small**	10	Orange	Orange	None	
15. Copper, brass, aluminum	None				None	

†Figures obtained with 12″ wheel on bench stand and are relative only. Actual length in each instance will vary with grinding wheel, pressure, etc. *Blue-white spurts. **Some wavy streaks.

Fig. 73-3 Spark streams of different metals. (Courtesy of Norton, Co., Worcester, MA)

straight-carbon tool bits or a worn file). By repeatedly spark testing and observing the spark streams, learn to compare the color, length of spark stream, and number of sprigs or stars given off by each grade of carbon steel tested. By the use of good judgment, one should be able to further classify other intermediate grades by the difference in the spark streams.

Illustrations of spark streams give an approximation of those produced by various metals and help to identify them, but they should not be used to replace the experience of actual spark testing.

Note that most of the illustrations show the sparks as spreading out in straight lines, but some indicate curving of the shafts. This curvature is one more important indication of a difference in the metal tested.

THE WELDER AND METALLURGY

Heat treating requires a wide knowledge of metals and alloys. This includes their reactions to heating and cooling cycles at various temperatures and air, oil, water, and molten salt baths. In short, heat treating is a highly specialized branch of the metal industry.

A welder should not be expected to be an expert in the field. However, awareness of some of the aspects of metallurgy, allows the welder to plan more intelligently and to carry out the procedures necessary to join most metals and alloys successfully. The welder should:

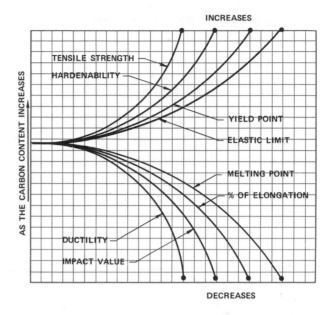

Fig. 73-4 Relation of carbon to physical characteristics

- be able to recognize what the material is.
- try to determine if the material has been heat treated and what steps, such as preheating, postheating, and cold-working, will be necessary to deliver the finished product in the desired condition.
- be aware of some of the chemical changes that take place when a weld is made. Even in a weld on carbon steel, the metal is almost instantly brought from room temperature to the molten state far above the Ac_3 point.
  Note: If the weld is progressing at a rapid rate, the weld and adjacent zone are severely quenched by the still-cold parent metal which has a much higher thermal conductivity than any other quenching metal.
- be aware of how much various metals and alloys expand or contract under the heat of fusion. Special attention should be given to the fact that the temperature achieved by the welding process is extremely high and the rate of heating and cooling is far higher than that attained by any other metalworking process.
- take advantage of the high temperature and high rate of heat transfer available in the variuos fusion-welding processes and use them to an advantage.
- In the event that the high temperature and high rate of heat transfer have a bad effect on the weldments, use every means available to counteract these negative effects. They can usually be minimized by the proper preheating or postheating operations.
- be aware that education in welding is a continuing lifetime process.

The low-carbon steels do not present many problems to the welder. As the carbon content rises through the medium- and high-carbon range, the problems multiply and become more complicated, especially if other alloying elements are introduced.

Regardless of the welder's knowledge of metals, it is wise to study and remember a few of the metallurgical constants shown in figure 73-4. This chart is not drawn to scale, and it is not intended to present accurately, the relation between the carbon content and the other characteristics shown. Its purpose is to show pictorially, the fact that these relationships do exist.

SOURCES OF WELDING INFORMATION

The American Welding Society publishes a trade journal and promotes interest in welding through technical meetings, welding shows, and the publication of handbooks and other technical data. This organization also cooperates with other engineering organizations and standards associations in the evaluation of welders, and the formulation of rules and regulations for welding procedures and for the testing of welds.

Many other engineering and trade journals devote considerable space to articles on welding. Other publications, such as *Welding Engineer* and *Welding Design and Fabrication,* publish authoritative, informative, and interesting articles on all phases of welding.

The American Society for Metals publishes the *Metals Handbook,* a very helpful, though not all-inclusive, treatment of the area.

Most of the manufacturers of welding equipment and supplies, as well as the manufacturers of primary metals and alloys will send, on request, brochures of their available handbooks and technical data.

Public libraries are always a valuable source of information.

REVIEW QUESTIONS

1. What should the first observation be when attempting to identify a metal?

2. Look up the specific gravity figures for several metals. Is specific gravity a good indication of what metal is being checked? Why?

3. How is the specific gravity of a metal obtained? How accurate is it?

4. Heat a small corner of a sheet of stainless steel and a piece of polished steel about the same gage to a red heat. Allow to cool and observe the results. What is the chief difference?

5. Place a few filings of zinc, lead, and tin on a steel sheet and apply heat to the bottom side of the sheet with an oxyacetylene flame until the metals oxidize and burn. What was observed about the color of the oxide?

6. Obtain some cadmium filings, or obtain a cadmium-plated article, such as a cadmium-plated screw, and heat it until it oxidizes. **CAUTION: It produces fumes which are highly poisonous.** Observe this oxide carefully. Learn to recognize it and avoid welding such material except under conditions of perfect ventilation. What is the color of cadmium oxide?

7. Obtain a chromium-bearing alloy, such as 18-8 stainless stellite or Stoody No. 1 and heat it until it oxidizes. What color does this oxide produce?

8. If a tungsten electrode is available, strike an arc and bring the electrode to incandescence. Allow it to cool and observe the color. Is there any similarity between the tungsten oxide and any of the others observed?

9. What is the appearance of the fracture in malleable iron?

10. Describe the results of a spark test on malleable iron.

11. The spark streams of both cast iron and high-carbon steel may emit sprigs or stars. What are the differences?

12. Name the common forms that spark streams are identified by.

13. Do these various components appear singly or in combinations?

14. What does the appearance of large amounts of sprigs or stars in a spark stream usually indicate?

15. What does a chip test indicate?

16. What is one of the fastest and most accurate methods of identifying metals and alloys?

17. Can all the elements be identified by the use of a spectrograph?

18. What is the relationship between the carbon content of steel and some of its other characteristics?

19. As the carbon content of steel and its alloys increases, does the metal become more difficult to weld?

SECTION 5
Employment
Opportunities in Welding

   In recent years, the metalworking industries have established a large
number of occupations which require skill and knowledge in welding. In
earlier years, welding occupations were concerned largely with repair work.
Today, the major employers of welding personnel are the manufacturing and
construction industries.

   Many opportunities for persons well trained in the various aspects of
welding and related fields exist in research and development, design, and the
operation of welding equipment. Other opportunities are found in the design,
construction, and maintenance of welding facilities. Small job shops constitute
another important source of employment. Skill and knowledge in welding
may be supplemental to another major occupation such as plumbing and
steamfitting, or aircraft maintenance.

   For all occupations related to the welding field, education and specialized
skills are necessary. Schools and the welding industry itself have been active
in providing such training. This training must be broad enough to make it
possible for skilled workers to adapt to new welding processes, conditions,
and applications.

UNIT 74 THE DEVELOPMENT AND APPLICATIONS
OF MODERN WELDING TECHNIQUES

The original welders were the skilled blacksmiths who forge-welded metals by heating and hammering.

Fusion welding came into being with the development of equipment to utilize the oxygen-hydrogen flame in the 1880s. About the same time research led to methods of producing acetylene gas cheaply along with equipment to handle this sensitive gas safely.

Electric arc welding was also developed in this era as well as the resistance welding process of which spot welding is the most common form.

The early use of all of these processes was limited to repair work. Actually, welding was a technique to be used when no other method of repair or maintenance worked.

The growth of the welding processes was slow and the number of people involved was relatively small even into the early 1920s. As more people in management and engineering became aware of its potentials, welding began to be used in a limited way as a production tool.

THE COATED ELECTRODE

Welding probably received its most significant stimulation with the development of the coated electrode. For the first time arc welders could produce welds which had physical characteristics equal to or better than the parent metal. This gave the engineers and designers the opportunity to develop new products which were lighter, stronger, and better looking than those produced by other metal-working processes. Industrial applications also led to a demand for many more welding operators, more and better welding equipment, and better materials and supplies. Welding research and education kept pace with the technical development.

Today, practically the entire metalworking field is contributing to research and improvement of the vastly expanded and improved welding processes. Colleges train scientists and engineers in the field of metallurgy and welding engineers. Industrial training programs and technical education programs train technicians to work between the engineering and production departments.

The training of welding operators up to the 1920s was primarily a function of a few private trade schools along with a few job shop operators who supplemented their income by teaching a limited number of men some of the skills. As the use of welding grew and the demand for welding operators increased, many public vocational schools offered courses in welding. Many manufacturers set up their own training centers.

The welding field has undergone considerable change and this change accounts for much of the increasing use of welding. Not only is welding equipment being constantly improved, but the price has been lowered in the face of rising costs in other fields. For instance, a modern arc-welding machine is vastly superior to one of the 1920s, but the cost of today's machine is less than one quarter of the comparable 1920 unit. Research and education have progressed so rapidly in the field that people associated with welding are engaged in one of the most comprehensive fields in existence.

MODERN APPLICATIONS

Welding applications range from space vehicles to rubbish cans and from the very thick shells of atomic reactors (12 inches to 16 inches) to the very thin liquid fuel tanks of rocket engines. Nearly the entire metalworking field employs at least some of the welding processes. It is estimated that our modern automobile would cost a thousand dollars or more over today's prices were it not for the extensive use of welding. Refrigerators, washing machines, and electric and gas stoves as well as gas- and oil-fired heating furnaces require welding in some form. Even cooking utensils employ spot or projection welding.

Many machine tool builders have turned to welding to provide their customers with equipment which is both stronger and more rigid. The builders of earth-moving equipment used in mining and higway construction combine a variety of metals to provide strength in areas of high stress as well as toughness or hardness to resist abrasion and wear.

Fig. 74-1   Pipe welding on the construction site

The transportation industry offers many jobs in the welding field. Most trucks, buses, subway trains, and railway passenger cars are welded to provide lighter, stronger units. Box cars and gondola cars are constructed of high-strength steels which are welded to provide greater carrying capacity. Even the rails are welded into mile-long sections to provide smoother, more economical operation with less chance of failure.

Most modern transport and military aircraft could not be made without the strength and economy of welding. The shipping industry is making freighters and tankers of huge size to remain competitive.

The highly automated petroleum industry depends on welding to provide the fractionating towers and piping that have to operate at high temperatures and high pressures for efficient production.

The cost and convenience of heating homes and powering industry and vehicles would be much different were it not for the thousands of miles of welded cross-country pipelines carrying petroleum and natural gas.

Gasoline, oil, and water storage tanks are rapidly and economically erected using welding.

Fig. 74-2 Welded bridge supports

Tanks of special alloys are welded to contain liquid oxygen, argon, nitrogen, and other liquefied gases at temperatures in the minus three hundred degree range.

The construction industry makes stronger and lighter steel-frame buildings by welding the members. Highway and railway bridges are commonly welded not only to provide a greater strength-to-weight ratio, but also in many cases, to provide structures which have a pleasing streamlined appearance as the one in figure 74-2. Our modern hydroelectric plants need welded penstocks to carry large volumes of water at high pressure to the turbines which are highly efficient due to welded steel construction.

Maintenance

The field of maintenance welding employs a large number of welding specialists who repair and replace equipment that has failed due to abrasion, corrosion, heat, or high stress. Most of these welders have studied the materials and methods required to provide for a variety of situations. For instance, in the field of hard facing there are combinations of abrasion, impact, chemical attack, and high heat. These specialists must have the knowledge and skill to choose the most effective materials and methods for the best results.

Producers and users of chemicals normally employ many welders in their maintenance department. They are called upon to weld, not only the common alloys, but also such metals as nickel, silver, copper, tantalum, and titanium.

The transportation industry uses welders to maintain such things as railroad rolling stock and the roads themselves. Truck fleets need welders to repair, replace, and sometimes strengthen dump truck bodies and the equipment used in transit-mix concrete service. Automobile repair shops require welders, especially in the body repair division. Airline repair facilities require highly skilled welders to repair and replace critical parts.

All four branches of the military use equipment which is of necessity subjected to the most severe abuse. They need maintenance personnel who can make repairs swiftly and effectively under far from ideal field welding conditions.

The mining, dredging, and earth-moving industries use huge costly machines which are constantly subjected to high stresses, high impact, severe abrasion, and, in some cases, severe corrosion. They must have welders with the experience and judgment necessary to keep the equipment in operating condition, usually with little supervision.

REVIEW QUESTIONS

1. What was the application of most early welding processes?

2. a. What development led to the increased use of welding as a production tool?
   b. Why did this new process encourage production welding?

3. Name three types of consumer goods in which welding plays an important part in manufacture.

4. Name two service industries which use welding to manufacture and maintain equipment.

5. Have costs in the welding field risen proportionately with costs in other manufacturing industries? To what can the increase or decrease in costs be attributed?

UNIT 75 OCCUPATIONS IN WELDING

OPERATOR'S ASSISTANTS

Occupations in the welding field may be said to offer unlimited opportunities. *Welder's helpers* are employed in some branches of the industry. Their function is to clean slag from welds and move or turn weldments for the welder. The welder's helper should have some general knowledge of what the welder does in order to be of useful assistance.

Tackers are the semiskilled workers who tack weld the various components of weldments. They must have some welding skill and be able and willing to follow the welder's directions.

Welding fitters work with the tackers. It is their responsibility to see that all the components are fitted accurately into their proper places and are tacked in such a manner that they will stay in proper alignment during the welding process. Weld fitting requires a knowledge of blueprints, the effects of expansion and contraction, the use of a variety of tools, plus the ability and desire to turn out first-class work with a minimum of supervision. Therefore, fitting is one of the better paying jobs in the welding shop.

WELDING OPERATORS

The *welding operator* does the actual welding. The operator may use manual processes or semiautomatic or automatic equipment. Pay varies with the welder's skill and knowledge, and the requirements of the job. Some operators work in situations where the ability to follow orders and run an acceptable bead in the flat position are the only requirements. Other welders are more highly skilled and are expected to produce acceptable beads or welds in the flat, horizontal, vertical, and overhead positions, figures 75-1 and 75-2. Their pay varies with the number of positions they can use.

In some industries the operators must pass a test to prove their ability. The most common test is the AWS, ASME minimum operator qualifying test which is given in any one or all four welding positions. Welders who qualify are paid at a higher rate than nonqualified personnel and in general, get a raise each time they qualify for a more difficult position.

Welders who must join special alloys that operate under high pressures

Fig. 75-1 A welding operator on a large production job.

or in high-heat service are tested by much more demanding tests under very rigid specifications. Naturally, their pay scale is even higher. Beyond this there are highly skilled welders with the ability and education to work in experimental laboratories with technicians and engineers on experimental and development work.

There is also the welder who, with sufficient skill and education, can take a blueprint and do all the necessary work to produce a part or a complicated machine. People of this type are in constant demand in prototype shops maintained in many industries. They are probably the highest paid workers and have some of the most challenging and interesting jobs.

Some welders with a high degree of skill and knowledge and at least a basic knowledge of small business management prefer to operate their own business. Most start rather modestly with not too large an investment in machinery and equipment and sometimes specialize in such fields as hard facing, repair, and maintenance. Gradually they branch out into subcontracting of manufactured items as they are able to increase their investment in capital equipment. Most of the successful job shop operators display more than an average amount of drive and determination.

WELDING SUPPORT TEAMS

The *welding support team* employs as many or more workers than the actual welding operations.

The highly educated scientist must have broad knowledge in the fields of chemistry, electricity, metallurgy, physics, and mathematics in order to direct teams of engineers in research, development, and design.

All the various engineers work individually or in teams to produce designs for new equipment, materials, and methods and usually have several technicians working with them. These technicians act in the capacity of communication officers between the engineering

Fig. 75-2 A welding operator performing under difficult conditions

department and the production shop. They conduct physical and chemical tests and may conduct welder qualifying tests. They use X-ray equipment, dye penetrant, magnetic flux testing equipment, and ultrasonic equipment used for physical testing of materials and welds. They require special education offered in community colleges and technical schools with special emphasis on mathematics, science, and mechanical drawing.

Direct layout persons and *template makers* are also most important to the support team. Their duties include laying out materials and marking them properly for subsequent operations. These operations include shearing, flame cutting, punching, flanging, rolling, and twisting. This work is either done directly on the material, or all the information is put on a template. A template may be metal, wood, or paper, depending on how durable it needs to be.

Both grades of layout persons command excellent pay with the direct layout person usually having a slight advantage.

Fig. 75-3 Flame-cutting operator using semiautomatic pipe cutter

The people who operate shears, brakes, punch presses, and rolls all have special knowledge and skills and are paid accordingly, but they usually receive less than the skilled welder or layout person.

Flame-cutting operators vary greatly in their skill and knowledge. Some operators of hand-held cutting torches are expected only to cut reasonably straight line cuts. Other burners are expected to cut difficult shapes and bevels, and they command a fairly high rate of pay. Some burners specialize in machine flame-cutting, figure 75-3. They may operate power-operated straight line cutting machines to sever or bevel sheets and plates. They must know how to choose the right burning tips, adjust to the right cutting pressures, and set the rate of travel for the most economical operation. In some cases they must also know what the width of the kerf or cut will be and compensate to produce parts of highly accurate dimensions.

Other specialists concentrate on operating expensive, highly sophisticated burning equipment which is either tape controlled or paper template controlled. This template may be a line drawing of the desired part or a cutout of the part placed on a contrasting background. It provides a pattern to be followed by an electric eye scanning device which activates the burning torch.

Many operators of such equipment are promoted from manual burners and operators of automatic straight line cutters. They have all the knowledge and skills of these operators plus the training to make adjustments in the electric devices of the shape-cutting machines.

Some of these operators use templates or patterns provided by the template-making department or drafting room. Other operators make their own patterns and are required to have a knowledge of exactly how the equipment reacts under varying conditions. These conditions include gas pressures and rates of travel in order to compensate for the width of the cut and for runout when the torch makes an abrupt change of direction.

Electronic and magnetic tracing machines produce highly accurate parts but only when the operator and template maker use all their knowledge and skill.

The compensation for flame cutters or burners varies with their skill and ability. In the case of operators of expensive shape cutting equipment, the pay is not only excellent, but the lack of highly trained personnel results in a great deal of overtime work.

TRAINING FOR WELDING OCCUPATIONS

There are many manufacturers today who will not hire anyone who does not have a high school education. Many schools are available to people who wish to specialize in the welding field. Most communities of any size have at least one public vocational-technical school with special courses in welding, both as a trade and in related fields. Requirements in mathematics, science, and mechanical drawing are designed to graduate welders with a background that can lead to upgrading. Several small communities may combine their resources to build and maintain area vocational schools with the same services and functions except that the vocational-technical students are bussed from their home schools for their shop and laboratory work.

Major manufacturers of welding equipment maintain their own schools to train welders for the convenience of their customers. Most of these manufacturers also offer short, intensive courses to help welders upgrade their skills. They also offer short courses designed to aid management, supervisors, technicians, and engineers to acquire a better understanding of welding processes and to develop more efficiency in the use of the tools and equipment available.

Community colleges and some private engineering schools offer courses which emphasize mathematics, science, engineering, drawing, and physical testing, usually on a two-year basis. Several colleges and universities offer specific degree programs in welding engineering. Many other colleges and engineering schools offer courses with more emphasis on welding technology with each succeeding year.

The working welder who desires a supervisory position has access to night school courses in mathematics, mechanical drawing, science, and management in public schools as well as advanced courses in community colleges and universities. There are also short, intensive-study courses in new techniques and equipment offered at a nominal cost by welding equipment manufacturers. The American Welding Society offers some excellent short courses each year, staffed by some of the most expert personnel in the various fields.

Practically all the manufacturers of welding equipment and supplies offer many technical manuals either on a free basis or on an at-cost basis. This information gives the welder the opportunity to learn by home study at practically no investment other than time.

Many of the large users of welding offer in-plant training to their employees in order that they may improve their skills and knowledge. Many of these same manufacturers today will pay all or part of their employees' tuition to private colleges as long as the courses they are taking lead to upgrading in their particular field of interest.

SUMMARY

In general, nearly all the people associated with the welding industry are paid at a rate higher than the industry-wide average in whatever field they are engaged. The rapid growth of welding technology offers unlimited opportunities for advancement limited only by the individual's abilities and ambitions. Educational opportunities are available at all levels and in practically all areas, both geographically and educationally.

REVIEW QUESTIONS

1. a. What are the duties of a welding fitter?
   b. What knowledge and skills does the welding fitter need?

2. What is the most common test given to welding operators?

3. What does the direct layout person do?

4. a. Name two methods by which automated burning equipment is controlled for flame cutting.
   b. What is the most important factor affecting the accuracy of this equipment?

5. Name at least four ways in which the welder can gain the skills and knowledge necessary for promotion.

6. Has the demand for skilled labor in the welding field increased or decreased over the years? Give reasons for your answer.

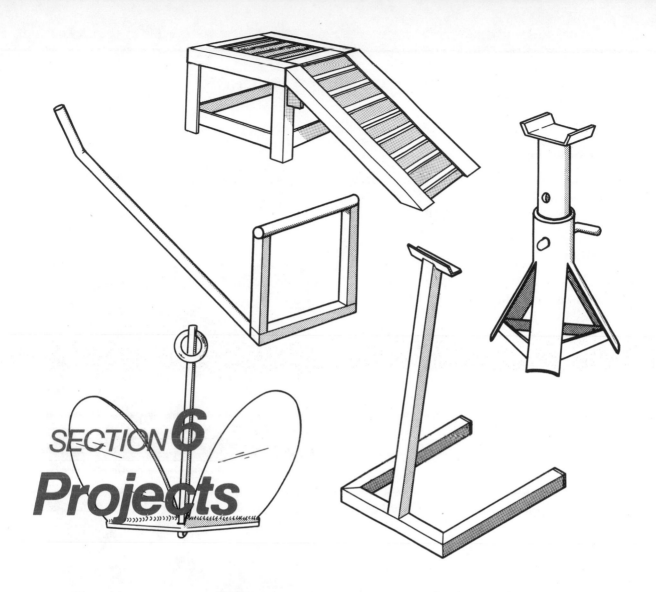

SECTION 6
Projects

The following projects are some of those developed over the years from various sources and were selected on the basis of utility, ease, or difficulty of production, and student interest. They offer the student the opportunity to put to practical use some of the knowledge and skills learned in the various welding, brazing, and flame-cutting units.

Completion and inspection of some or all of the projects help to indicate the degree of skill and knowledge that has been acquired as well as pointing out areas of lack of skill and knowledge.

Additional sources for projects include *School Shop* and several other magazines. Several companies offer project books at a price below their actual printing cost:

Lincoln Electric Company, Cleveland, Ohio, offers *Arc Welding Projects for the School Shop and Farm.*

Di-Acro, Oneill-Irwin, Lake City, Minn., offers *Projects in General Metal Working.* Many of these projects can be adapted to welded designs.

Hobart Welding School, Troy, Ohio, offers *Welding Projects for School Shops.*

PROJECT 1   BOAT ANCHOR

Suggestions for Design

The size and thickness of the parts can be increased to provide a heavier anchor. The anchor shown works well for a 14-ft. boat. Pointed flukes can be used, but they tend to gouge the bottom of the boat.

Materials

Nos. 1, 2, and 3 from 3/8-in. steel plate, as indicated in the drawing, or thicker if desired.

No. 4 — 3/16-in. steel or bronze welding rod formed into a ring 1 in. to 1 1/2 in. I.D.

No. 5 — 3/8-in. diameter rod, 2 1/2 in. long E-6024 electrodes.

Procedure

1. Lay out base flukes and tongue on individual steel plates, or nested on one plate as in the photo.

2. Centerpunch all layout lines lightly and flame-cut.

3. Grind all flame-cut edges for a smooth surface. Draw-file all exposed edges to prevent a slightly rounded surface that will not scratch or damage the bottom of the boat.

4. Place the flukes on the centerline of the base and tack-weld, making sure the parts are aligned and square with each other.

5. Double check the tacked assembly to make sure the two flukes are in the same plane and square with the base. Weld with E-6024 electrodes, observing all the precautions you have learned when practice welding with this type of electrode.

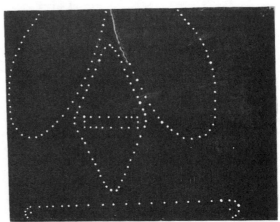

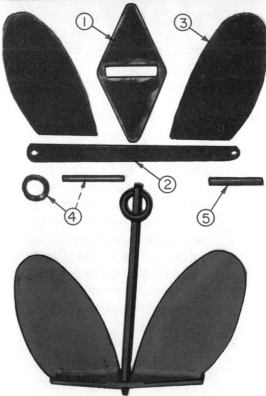

1 – BASE   2 – TONGUE   3 – FLUKE
4 – RING   5 – ROD

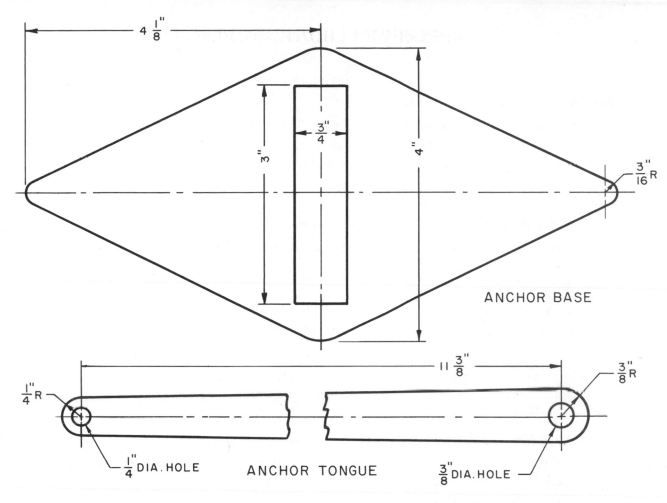

ANCHOR BASE

ANCHOR TONGUE

6. Form the ring by wrapping 3/16-in. rod around a 1-in. diameter shaft. Insert in the tongue, line up ends, and weld or braze.

7. Place the tongue in the anchor as in the photo and insert 3/8-in. x 2 1/2-in. rod through the hole provided. Align the rod on centerlines, tack, and weld.

8. Thoroughly clean all slag and splatter from the assembly. Prime and paint. The choice of color depends on individual taste. However, aluminum or copper seem to reflect light rather well and help in the event the anchor rope breaks and it is necessary to search for the anchor.

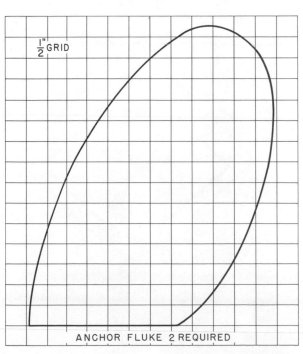

ANCHOR FLUKE 2 REQUIRED

PROJECT 2   WALL HUNG FLOWER POT

Suggestions for Design

The key to ornamental work is proportion. If a larger flowerpot is to be used, the size of the frame would have to be altered in keeping with the size of the ring required. Also, the diameter of the rods may have to be changed to present a pleasing appearance.

Materials

No. 1 — 3/16-in. diameter steel rods for the frame

No. 2 — 3/4-in. x 1/8-in. steel flat bar for the ring and bracket

Procedure

1. Cut 3/16-in. rods to the length indicated to form both square frames.

2. Determine the length of the corner rods as indicated in the schematic and cut them to length.

3. Place the 4 rods on a flat surface and tack them to form the outer frame. Also tack the 4 rods that form the inner frame, making sure they are square before tacking.

4. Check for squareness by measuring the diagonals. If one diagonal is longer, pull the corners together until the two diagonals are equal.

5. Weld the frames.

   Note: With most weldments, each assembly and subassembly should be completely tacked and checked before any welding is attempted. Trying to weld one piece at a time usually results in a distorted product.

6. Place the large frame on a flat surface and hold it in place with steel spacers 1 1/2 in. thick.

7. Place the smaller frame on top of the spacers and center this with the larger frame.

8. Hold the corner rods in place and tack both ends of all four.

9. Remove the spacers and weld.

10. Cut the 3/4-in. x 1/8-in. flat stock to length and roll to dimensions indicated.

    Note: If the dimensions are to be accurate, you will have to consider the thickness of the material used. In this case the neutral diameter will be the inside diameter plus 1/8 inch and the neutral, or true length of the bar will be the product of the neutral diameter times 3.1416.

11. Cut the 3/4-in. x 1/8-in. bracket to the dimensions indicated and weld to the ring.

12. Place this assembly on the frame. Tack and check for squareness and alignment.

13. Clean, prime, and paint.

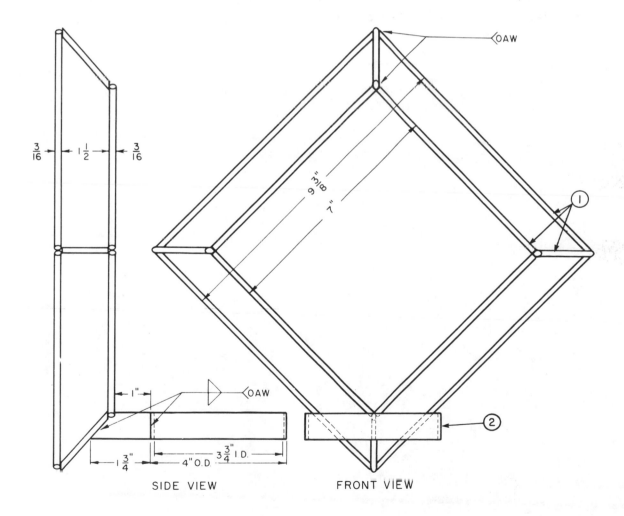

SIDE VIEW   FRONT VIEW

PROJECT 3   EASY-STORING, SELF-ALIGNING TRESTLE

Suggestions for Design

The dimensions indicated can be altered to suit specific needs.

Materials

CROSSBARS AND LEGS — 1/2-in. standard black iron pipe

SOCKETS — 1-in. standard black-iron pipe 5 inches long

STEEL PLATES — as indicated

TENSION RODS — 3/16-in. diameter gas welding rod

This self-aligning trestle is a safe, strong, convenient support for use about the home or on construction jobs. It can be readily assembled and adjusted to meet various conditions. Another feature is the ease with which it can be disassembled for storage or transportation.

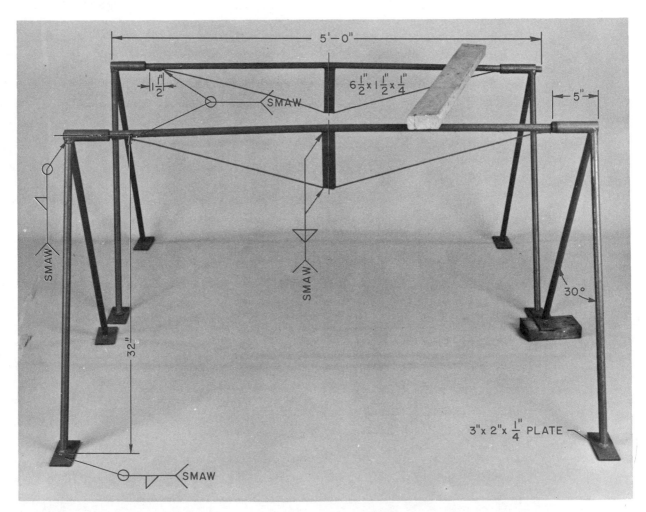

A variety of trestles can be designed and constructed to meet specific job requirements. When a new design is conceived, consideration should be given to the size of the components in relation to the weight and distribution of the load.

Procedure

1. Lay out and cut 1/2-in. pipe for legs.

2. Align with 1-in. x 5-in. socket, tack and weld as indicated.

3. Cut pipe for the crossbar.

4. Lay out and bend 3/16-in. welding rod to fit crossbar as indicated.

   Note: The distance from the rod to the crossbar at the center should be 1/2 in. less than the length of the vertical plate.

5. Weld the rod to the crossbar at each end. Attach the 6 1/2-in. x 1 1/2-in. x 1/4-in. bar to the rod as indicated.

6. Heat the rod evenly throughout its length until the vertical plate just fits between the pipe and rod.

7. Tack and weld as rapidly as possible.

   Note: When the tension rod cools, the pipe will have approximately 1/2-in. camber.

8. Insert the crossbar into sockets. Place 3-in. x 2-in. plates under the legs. Tack and weld on a flat surface as indicated.

PROJECT 4 MOTORCYCLE STAND

Suggestions for Design

This stand is highly versatile, but may require dimensions other than those indicated to properly fit a given motorcycle. The most important of these variations is in the height of the stand.

Materials

Nos. 1, 2, and 3 — 1-in. square mild steel

Nos. 4 and 5 — 1-in. round mild steel

Procedure

1. Lay out and cut 1-in. square stock for base and supports.

2. Arc weld these three pieces in place.

3. Grind or file a 1-in. radius curvature in the top of the supports.

4. Arc weld the 1-in. round stock in place.

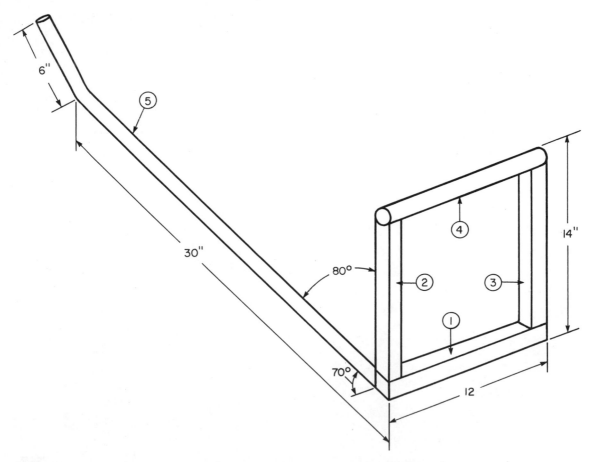

5. Heat the 36-in. piece of 1-in. round stock 6 inches from the end with a large sized oxyacetylene tip and bend a handle at this point.

6. Grind or file a 20-degree angle on one side of the base.

7. Arc weld the lever arm to the base at the 110-degree angle. The supports must also be at an 80-degree angle as shown.

8. Remove all slag, rust, and scale.

9. Prime and paint.

PROJECT 5   SELF-CLOSING MAILBOX

Suggestions for Design

   The suggested design has been carefully tested and works well. Any deviation from the drawing may result in a product that is unworkable. However, 18- or 20-gage steel can be substituted for the 16-gage as indicated, without affecting the workability or weldability of the unit.

Materials

   No. 1 — 13 9/16-in. x 16-in. x
    16-gage steel.

   No. 2 — 11 5/16-in. x 16 3/4-in. x
    16-gage steel.

   No. 3 — 5-in. x 7-in. x 16-gage steel.

   No. 4 — 5-in. x 7-in. x 16-gage steel.

   No. 5 — 3/4-in. x 1 1/2-in. x 1/8-in.
    steel

   No. 6 — 3/8-in. x 1 7/16-in. x
    16-gage steel

   No. 7 — 9/16-in. diameter brass x
    1 in.

   No. 8 — 3/16-in. x 3/4-in. diameter
    brass

   No. 9 — 1/4-in. lockwasher

   No.10 — 1/4-in. –20 hex nut

Procedure

1. Accurately lay out, cut, and drill pieces Nos. 3 and 4. They can then be used as templates while forming pieces Nos. 1 and 2. Piece No. 2 will probably have to be bent in a brake to form the radii indicated.

2. Assemble and tack weld pieces Nos. 1 and 3 as well as pieces Nos. 2 and 4. Use the same procedure as when tack welding any assembly. For example, tack progressively from one end.

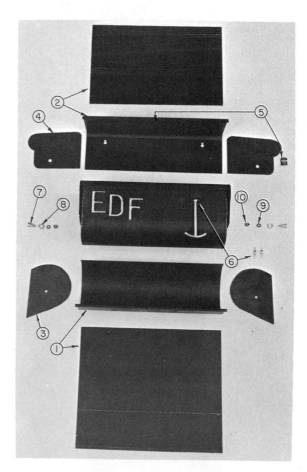

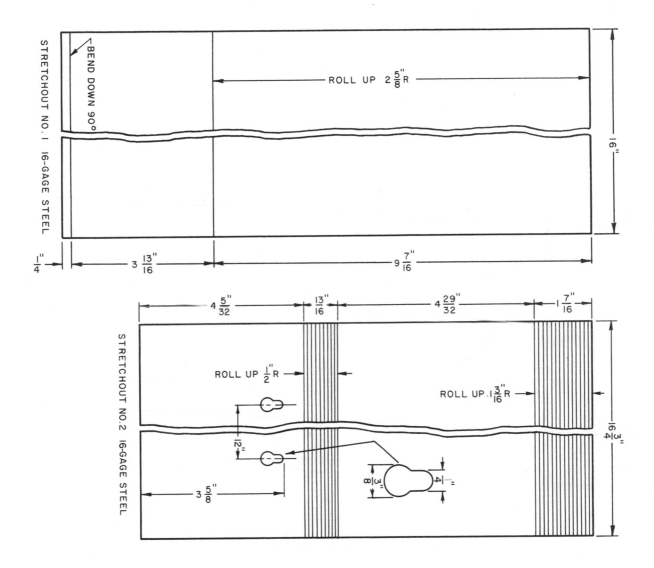

3. Make sure the fitup is reasonably tight and weld both assemblies, using skills learned when making outside corner welds.

4. Wire brush all welds, both outside and inside, to remove all scale that would interface with the adherence of the paint.

5. Form No. 5 and attach to center of No. 2 as per drawing. Braze welding will probably do the neatest job.

6. Form clips No. 6 and save until anchor is attached.

7. Make the initials from 1/4-in. bronze welding rod cut to size and assembled, using No. 3 Easy-Flo® alloy. This alloy gives a good color match and tends to build up to produce smooth joints. The initials shown are 2 3/4 inches high x 1 1/2 inches wide. Some letters such as I, M, W, will vary somewhat in width.

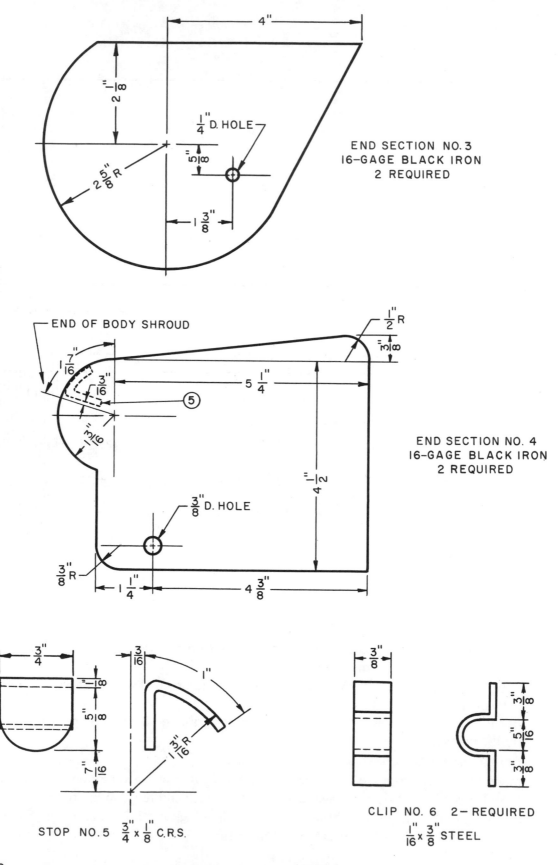

END SECTION NO. 3
16-GAGE BLACK IRON
2 REQUIRED

END OF BODY SHROUD

END SECTION NO. 4
16-GAGE BLACK IRON
2 REQUIRED

STOP NO. 5 $\frac{3}{4}$" x $\frac{1}{8}$" C.R.S.

CLIP NO. 6   2 – REQUIRED
$\frac{1}{16}$" x $\frac{3}{8}$" STEEL

8. Make the anchor from a 1-in. long top bar, a 5-in. long vertical bar, and a curved section made from 2 3/4-in. long rod bent to shape. Assemble and braze these pieces with No. 3 Easy-Flo®.

9. Carefully buff both the initials and anchor. The letters can be attached to the steel by soft soldering with a small 4x or 5x flame.

10. Attach the clips (No. 6) by resistance welding, if available. Otherwise, silver braze with No. 3 Easy-Flo® for strength.

11. Remove all residual flux and rebuff the initials if necessary. Carefully cover initials and anchor with masking tape.

12. Prime and paint. Black paint makes a good contrast with the highly polished brass. Once the paint has dried and has been inspected for satisfactory texture, strip the masking tape off and, if necessary, clean the surface with a solvent.

13. Coat the brass with a clear lacquer. Clear fingernail polish works well and a small applicator brush is attached to the cap. The pivots, No. 7, should also be coated.

14. If the pivots shown cannot be made, 1/4-in. x 3/4-in. hex headed bolts can be substituted. In this case, a nut will be required on each side of piece No. 3. Piece No. 4 will need to be drilled 1/4 inch in diameter instead of 3/8 inch. Spacer No. 8 can be eliminated without seriously affecting the operation of the unit.

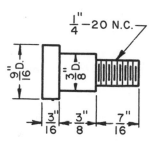

PIVOT NO. 7 2-REQUIRED
$\frac{9}{16}$" DIA. BRASS

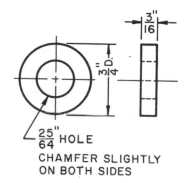

$\frac{25}{64}$" HOLE
CHAMFER SLIGHTLY
ON BOTH SIDES

SPACER NO. 8 2-REQUIRED
$\frac{3}{4}$" DIA. BRASS

PROJECT 6   AUTOMOBILE RAMP

Suggestions for Design

These ramps have been designed to fit most cars. In some cases automobiles with very low frames may require a longer ramp. This may be necessary so that the wheel will reach the lower part of the ramp before the body comes in contact with the upper platform.

Materials

Nos. 1, 2, and 3 — 1/4-in. x 1 1/4-in. x 1 1/4-in. angle iron

Nos. 4, 5, and 6 — 1/4-in. x 1-in. bar stock

Procedure

1. Select material, lay out, and cut to proper size with power hacksaw.

2. Lay out pieces for coped joint on stand top and cut with hand torch. Grind and file for proper fit.

3. Tack weld stand top, making sure pieces are in plane and square.

4. Weld stand top 100 percent.

5. Locate stand crossbars. Tack weld and weld both sides on vertical joint.

6. Locate stand legs on stand and tack on heel and toes of angle.

7. Locate leg braces on legs, square legs, tack weld braces, check unit for square, weld legs to stand 100 percent, and weld braces to legs.

8. Locate ramp crossbars. Tack weld in position, making sure unit is square. Weld crossbars both sides on vertical joint.

9. Fabricate ramp holding lugs from 1/4-in. x 1-in. bar stock. Locate on stand and tack weld.

10. Place ramp on lugs to ensure proper alignment. Weld lugs to stand on both sides of vertical joint.

11. File rough spots or rough edges smooth.

12. Prime and paint.

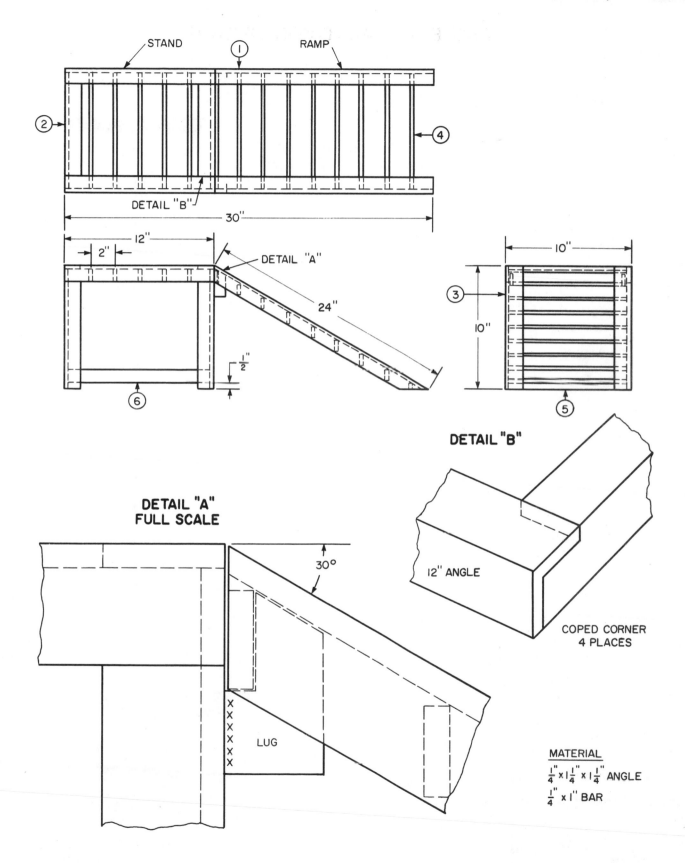

STAND   ①   RAMP

②

④

DETAIL "B"

30"

12"

2"

DETAIL "A"

24"

10"

10"

③

⑤

½"

⑥

DETAIL "B"

12" ANGLE

COPED CORNER
4 PLACES

**DETAIL "A"
FULL SCALE**

30°

LUG

MATERIAL
$\frac{1}{4}$" x $1\frac{1}{4}$" x $1\frac{1}{4}$" ANGLE
$\frac{1}{4}$" x 1" BAR

275

PROJECT 7   HOLE DIGGER OR SPUD

Suggestions for Design

A hole digger of the dimensions and materials indicated works well for planting shrubs, trees, vegetables, and flower plants started in peat moss pots which may require fairly deep holes.

Materials

No. 1 — 4-in. x 10-in. x 3/8-in. steel plate
No. 2 — 4-in. x 1 1/2-in. x 3/8-in. steel plate
No. 3 — 1/2-in. black iron pipe 3 ft. 1 in. long
No. 4 — 1/2-in. black iron pipe 8-in. long

Procedure

1. Cut plate No. 1 and notch as indicated to take 1/2-in. pipe. Make the blade by milling, shaping, or grinding.

2. Cut piece No. 2. For 100 percent fusion, this piece should be beveled from both sides on the broad end.

3. Place both Nos. 1 and 2 on a flat surface, align, and tack weld.

4. Check for alignment and weld with a 5/32-in. E-6010 or E-6012 electrode.

5. Insert No. 3 in No. 1 and carefully align the centerline of the pipe and plate in both directions before tacking. The procedure can be simplified if No. 1 is placed on a 1/4-in. plate to properly space the plate and pipe.

6. Weld or braze No. 3 to No. 1.

7. Contour the top end of No. 3 by grinding or filing to fit No. 4.

8. Align and tack the tee handle in place.

9. Weld or braze the tee handle.

10. Thoroughly clean off all scale, slag, and flux.

11. Prime and paint with some color compatible with garden tools.

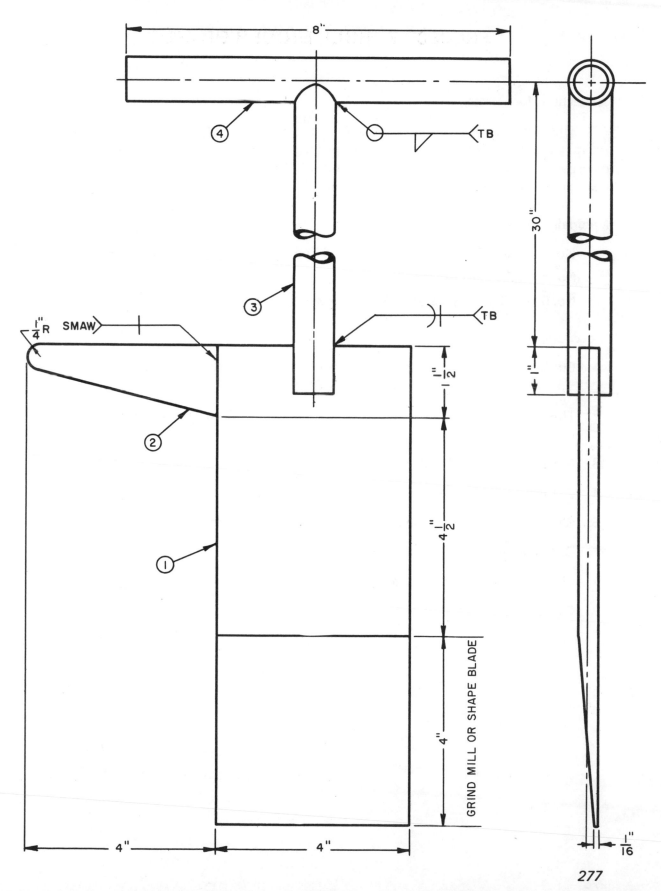

8"

TB

④

③

TB

$\frac{1}{2}$"

$\frac{1}{4}$"R SMAW

②

①

$4\frac{1}{2}$"

30"

1"

GRIND MILL OR SHAPE BLADE

4"

4" 4"

$\frac{1}{16}$"

277

PROJECT 8   LAWN SIGN

Suggestions for Design

This sign is designed to be secured to a post or other elevating environment. This project may also be constructed to be self-supporting by elongating the side support. This then may be sharpened to a point for easy ground installation. It is also possible that longer names may require a slightly enlarged project.

Materials

Nos. 1, 2, and 3 — 1/8-in. x 1/2-in. band iron

No. 4 — 22 ga. galvanized sheet

No. 5 — 1/8-in. oxyacetylene welding rod

Procedure

1. Cut an 18-in. piece of 1/8-in. x 1/2-in. band iron and bend it to 90 degrees, 4 inches from one end.

2. This sign support is then oxyacetylene welded to a 13-in. side support 4 inches from the bottom.

3. Cut a 26-in. piece of 1/8-in. x 1/2-in. band iron and bend the scrolls.

4. Cut two 6-in. pieces of the same size band iron and bend into small scrolls.

5. Check all pieces for fit and assemble by oxyacetylene welding.

6. Drill holes in band iron as indicated.

7. Cut 22 ga. galvanized sheet to shape and drill or punch holes as indicated.

8. Construct S hooks from 1/8-in. oxyacetylene welding rod.

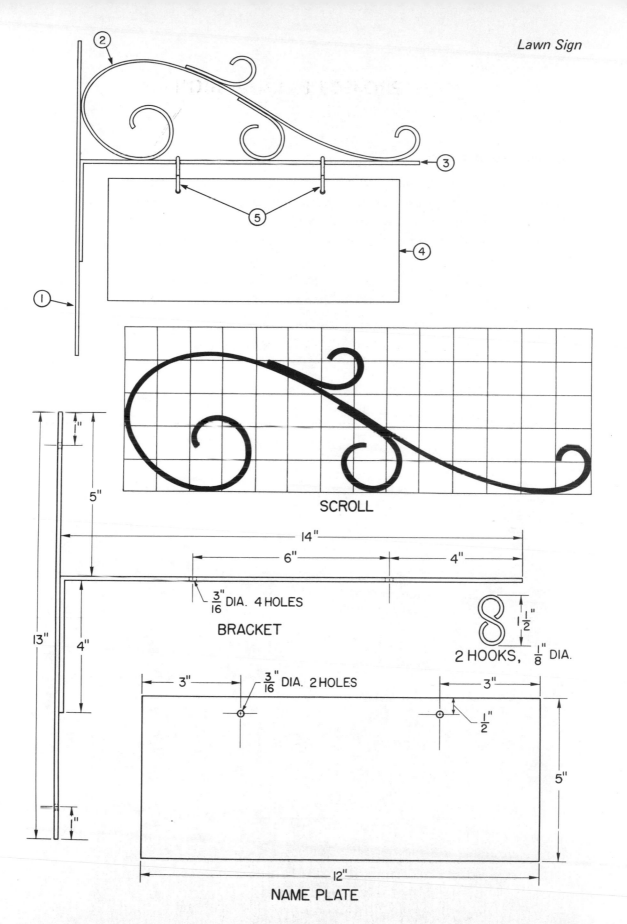

SCROLL

BRACKET

$\frac{3}{16}$" DIA. 4 HOLES

$1\frac{1}{2}$"

2 HOOKS, $\frac{1}{8}$" DIA.

$\frac{3}{16}$" DIA. 2 HOLES

$\frac{1}{2}$"

NAME PLATE

PROJECT 9   HOLD DOWN DEVICE

Suggestions for Design

   The device shown is large enough to be serviceable for most jobs. The primary purpose is to hold small, light pieces in alignment during welding, brazing, and tacking operations, especially where the flame force might be great enough to move workpieces such as the round pieces shown in the photo. These devices can be made in a variety of sizes to suit special needs.

   There may be cases when the rounded nose of No. 2 may be more efficient if it is ground to a point instead of the radius shown.

Materials

   No. 1 — 1 1/2-in. diameter x 8-in. long steel shaft

   No. 2 — 3/16-in. x 1 3/8-in. x 4 3/8-in. steel plate

   No. 3 — 3/16-in. x 4-in. x 4 3/8-in. steel plate

Procedure

1. Cut shaft No. 1 to length.

2. Lay out and cut 3/16-in. plates Nos. 2 and 3. Clean and deburr both pieces.

3. Set up pieces Nos. 1 and 2 on a flat surface and tack weld, making sure they are square with each other.

4. Stand this assembly so that No. 2 bears on the flat surface and insert No. 3 under opposite end of shaft No. 1. Make sure Nos. 1 and 3 are flush with each other and Nos. 2 and 3 are parallel.

5. Tack Nos. 1 and 3. Check again to be sure that all three pieces are properly aligned.

6. Weld as indicated in the drawing, using either a 1/8-in. or 5.32-in. diameter E-6012 electrode.

7. Remove all slag, rust, and scale.

8. Prime and paint.

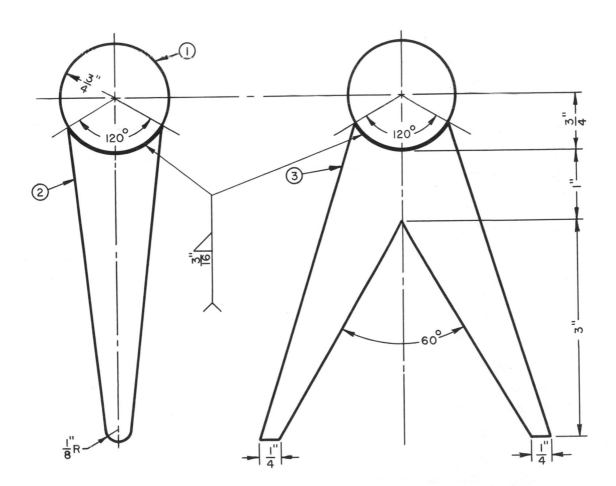

PROJECT 10   AUTOMOBILE JACK STAND

Suggestions for Design

These stands have been designed to be heavy and high enough to support an automobile. In the case of use with an especially heavy vehicle, a stronger material may be necessary. A small chain used to keep the pin permanently attached to the stand is also a worthwhile addition.

Materials

No. 1 — 2-in. standard pipe

No. 2 — 1 1/2-in. standard pipe

No. 3 — 1/4-in. x 1-in. bar stock

No. 4 — 1/4-in. x 2-in. bar stock

No. 5 — 1/2-in. round cold rolled steel

Procedure

1. Select material, lay out to size, and machine saw to proper length, 2 posts and 2 stands.

2. Select material for post pad, lay out bend lines, heat with a large size welding tip, and bend to fixture angle.

3. Lay out adjustment holes on centerline of post and drill 1/2-in. holes.

4. Lay out cut-lines and hole on stand and drill hole on one side. Using post as fixture inside stand, drill second side across diameter. Double check all sets of holes.

5. Set up stand on table and cut 3 leg lines with an oxyacetylene cutting torch.

6. Using a medium size welding tip, bend 3 legs to proper fixture angle.

7. Using cold chisel or pneumatic chisel, clean slag from stand legs.

8. Select material for 3 stand brace bars and cut to size with hand hacksaw.

9. Locate braces on stand and tack weld using 1/8-in. E-6010 or E-6011 electrode.

10. After checking for proper fitup, weld braces with same electrode.

11. Locate post pads on posts and tack weld with same electrode. Check fitup and weld pad to post using 1/8-in. E-7018 electrode.

12. Select material for pin, hand hacksaw to size. Using a medium size welding tip, bend to proper fixture angle.

13. Clean all parts and file all sharp edges.

14. Assemble both units and check all adjustments.

15. Prime and paint.

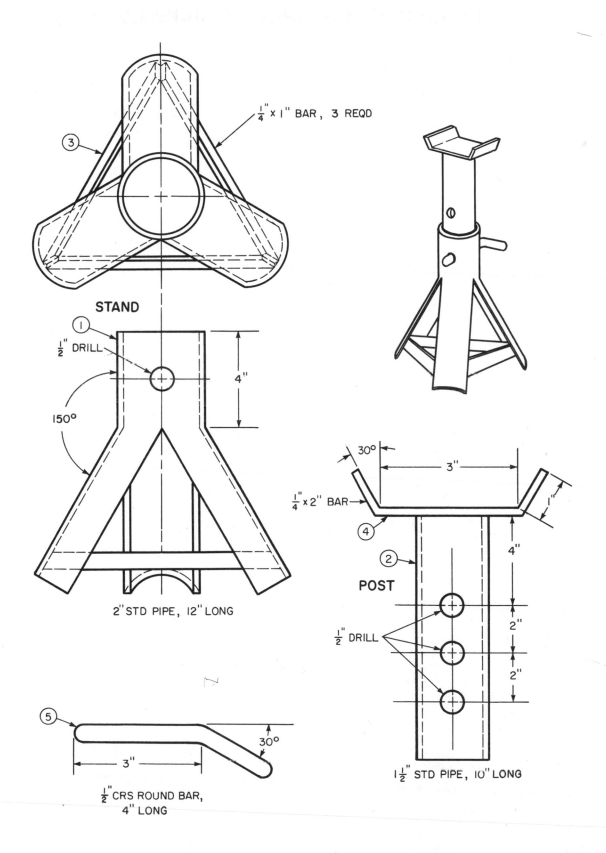

③ $\frac{1}{4}$" x 1" BAR, 3 REQD

STAND

① $\frac{1}{2}$" DRILL

4"

150°

2" STD PIPE, 12" LONG

30°

3"

$\frac{1}{4}$" x 2" BAR

④

② **POST**

$\frac{1}{2}$" DRILL

1"

4"

2"

2"

$1\frac{1}{2}$" STD PIPE, 10" LONG

⑤

30°

3"

$\frac{1}{2}$" CRS ROUND BAR, 4" LONG

PROJECT 11   SNOWMOBILE STAND

Suggestions for Design

The major design necessity for this project is stability. The angle of the support and the size of the base are of great importance. In some cases, it may be considered appropriate to include two supports.

Materials

Nos. 1, 2, 3, and 4 — 1-in. square tubing

No. 5 — 1/8-in. x 1-in. x 1-in. angle iron

Nos. 6 and 7 — 1-in. x 1-in. x 1/8-in. plate

Procedure

1. Select 1-in. square tubing for job.

2. Lay out and cut to size with hand hacksaw.

3. Lay out angles, grind and file to size.

4. Tack weld three base pieces together using the oxyacetylene process.

5. Tack weld two cap pieces on base.

6. Making sure base is square and in plane, weld all joints.

7. There is to be no grinding on any joints.

8. Locate angle iron in post. Tack and weld.

9. Locate post on base in proper position and weld.

10. After all welding is completed, finish any rough spots with a file.

11. Prime and paint.

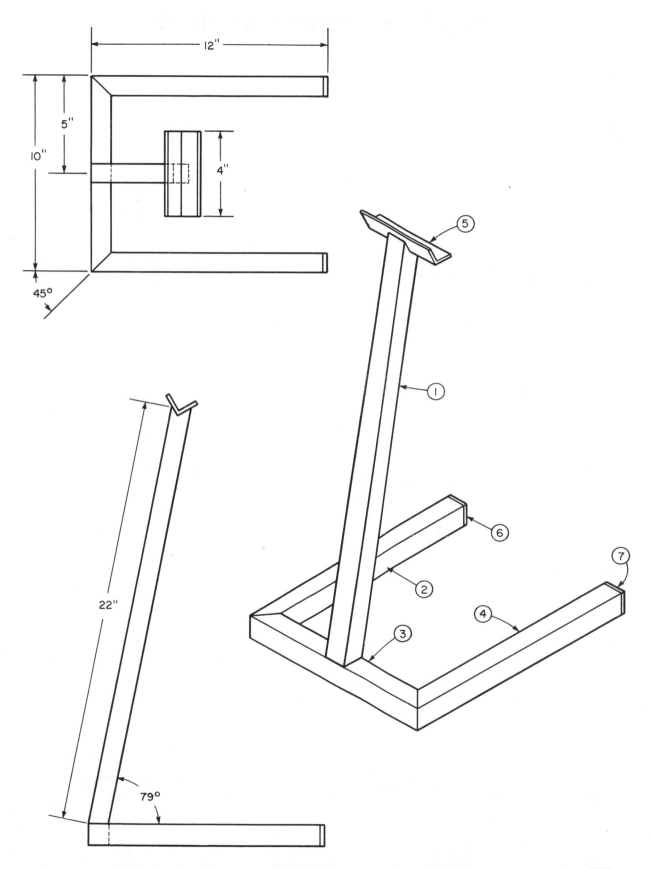

PROJECT 12   CHRISTMAS TREE STAND

Suggestions for Design

This tree stand is designed for heavy duty use. In cases which require light use, the dimensions of the legs and the pipe may be reduced.

Materials

No. 1 — 6-in. standard pipe

No. 2 — 1/4-in. x 1-in. bar stock

No. 3 — 1/4-in. steel plate

No. 4 — 3/8-in. 16 N.C. nuts

Procedure

1. Saw cut one piece of 6-in. pipe 6 inches long.
2. Using dividers lay out for I.D. of 6-in. pipe on 1/4-in. plate and oxyacetylene-torch cut.
3. Cut 3 pieces of 1/4-in. x 1-in. bar stock approximately 18 inches long.
4. Form bar stock as shown, either in bender or around a piece of 3/4-in. pipe.
5. Measure the circumference of a 6-in. pipe and lay out for leg location and holes.
6. Drill holes as shown at least 7/16 inch.
7. Bevel one edge of 1/4-in. plate as shown in detail "A".
8. Tack weld 1/4-in. plate with bevel to outside with 1/8-in. electrode in 4 places. Make sure the plate is flush with the bottom edge of the pipe.
9. Weld 1/4-in. plate to pipe from outside only, bringing weld up flush.
10. Tack weld three 3/8-in. 16 N.C. nuts over holes in pipe. Make sure nuts are centered over holes.
11. Enlarge two tack welds on each nut. Nuts do not have to be solid welded.
12. Locate legs on pipe and tack weld into position. Check. Set up.
13. Weld legs to pipe using 1/8-in. electrode in 3 places designated by x.
14. Clean all slag from welds with a chisel and brush.
15. Paint desired color.
16. Insert three 3/8-in. 16 N.C. x 3-in. long wing bolts into nuts to hold tree.

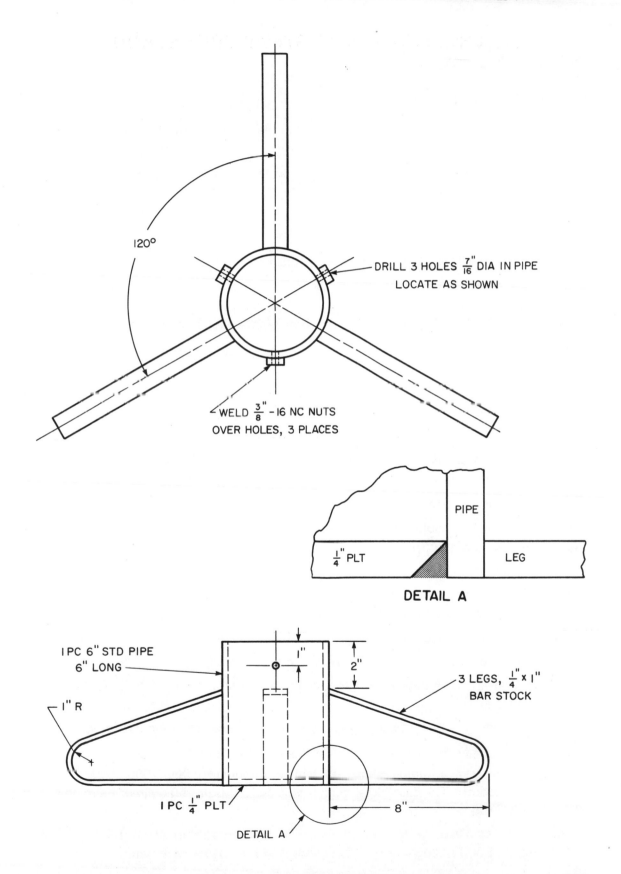

GLOSSARY

Acetylene: A gas formed from the chemical compounds of carbon and hydrogen.

Acetylene Feather: The light blue streamers in the flame as a result of an excess amount of acetylene.

Alloy: A mixture of two or more elements, at least one of which is a metal, which may have different characteristics than either of the base elements.

Alloy Steel: A form of steel containing a relatively high percentage of manganese, nickel, molybdenum, or chrome.

Alternating Current (AC): The flow of electrons that reverses itself continuously.

Ammonium Persulphate: An oxidizing agent used to etch metals.

Amperage: The unit of measure used to vary the amount of current applied to the arc welding circuit.

Annealing: The process used to soften a metal by heating it and then letting it cool slowly at a predetermined rate.

Arc: The flow of electricity through a gas which resists this passage. It is this resistance which creates the high temperatures needed for arc welding.

Arc Blow: The movement of an arc from its expected path due to magnetic forces.

Arc Welding: A welding process in which the parent metal is fused as a result of an arc produced by the flow of current through the atmosphere.

Argon: A colorless, odorless, inert gas used in TIG and MIG welding.

Austenite: The solid solution of carbon and iron.

Btu (British Thermal Unit): The quantity of heat required to raise the temperature of one pound of water one degree fahrenheit.

Backfire: A "popping" sound produced by the torch when improper use blows out the flame, or the boiling and exploding of a small piece of metal when severely overheated.

Backhand Welding: Welding in the opposite direction in which the torch is pointed.

Bead: The fusion area or the amount of filler metal deposited when welding.

Bessemer Converter: A large pear-shaped tank used to refine iron by forcing cold air through molten iron ore.

Bevel: The angle at which the edge of a piece of metal is cut to prepare it for welding.

Blast Furnace: A large stack used to refine pig iron from iron ore.

Brazing: A process of joining metals at a temperature greater than 1000 degrees Fahrenheit by capillary action. The base metal must have a melting point 50 degrees Fahrenheit higher than the filler rod.

Brinell Hardness Test: A test which measures the penetration of an 1/8" diameter ball under a load of 3000 kilograms for 10 seconds into the surface of a material.

Bursting Disc: A safety device which allows gas to escape from a cylinder which has been subjected to undue heat.

Butt Weld: A convex bead used to join two flat edges which may contact each other.

Cable: A heavy wire used to conduct current from its source to the work and then to the ground.

Carbon: A nonmetallic element added to steel to increase its hardness.

Carbon Steel: A form of steel which only contains carbon and no other alloying element.

Case-Hardening: The hardening of steel by adding carbon to its surface.

Carburizing Flame: The resulting flame which contains an excess amount of acetylene. This flame can be identified by its elongated acetylene feather.

Cementite: A compound of iron and carbon. Also known as iron carbide.

Charpy Test: An impact test in which both ends of the specimen is held.

Chip Testing: A method of identifying steel by comparing its resistance to being sheared with a known material.

Circuit: The path outlined by the flow of electricity from its source to a ground.

Coke: An organic fuel derived from the baking of coal.

Collet: A circular clamp used to hold the TIG electrode securely in the torch.

Color Testing: A method of identifying steel by the color of the arc given off when being subjected to a welding arc.

Compression Strength: The amount of squeezing force a material can withstand before failure.

Conductivity: The transfer of one form of energy from one place to another.

Cored-Wire Welding: A MIG welding process using intense heat, high deposition rate, and a flux cored wire for carbon steel welding.

Corner Weld: A triangular bead used on the inside or outside of two pieces of metal at a 90 degree angle and forming a shape resembling the letter L.

Crater: The concave area of a working weld bead which is formed from the pressures which surround it.

Critical Temperature: The temperature at which a magnetic material becomes nonmagnetic and subject to vast molecular changes.

Cross Section: The amount and shape of an area exposed when an object is cut on a certain plane.

Cutting Torch: A special torch with many orifices in the tip. The orifices on the outside of the tip are preheat holes. The large center hole delivers high-pressure oxygen when the trigger on the torch is pulled.

Cyaniding: A method of case-hardening steel by heating it in the presence of a cyanide salt and then quenching it.

Cylinder: A portable container used to store gasses under pressure.

Decarburization: The removal of carbon from the surface of steel.

Diamond Brale: A conical diamond point used in the Rockwell hardness test.

Diaphragm: A flexible seal which allows the regulator valve to open, yet does not let the pressurized gas escape into the atmosphere.

Diffraction Grating: A polished surface with parallel lines that produces colors by separation. Used to identify metals by examining the various colors given off.

Direct Current (DC): The flow of electrons which continuously flow in one direction.

Ductile: The ability of a material to be drawn or hammered out.

Elastic Limit: The amount of pulling force at which any additional stress produces a permanent deformation.

Electrical Conductivity: The ability to transfer electrons from one area to another.

Electrode: The portion of the welding circuit which comes in closest contact with the parent metal and at which the arc originates.

Electrode Holder: A spring clamp used to hold the electrode and provide a handle for the welding operator.

Elongation: The amount of difference between original length of material and its length after being tensile tested.

Extensometer: An extensometer is a micrometer device used to measure the amount a specimen will stretch before reaching its elastic limit.

Fatigue Value: The amount of fluctuating stresses a material can withstand before breaking.

Feeder Roll: A mechanical device used to supply wire to the MIG gun at a specified rate.

Ferrite: Pure iron contained in plain carbon steel.

Filler Metal: Material used to increase and fill the amount of material incorporated into the weld.

Fillet Weld: A triangular bead used to join two pieces of metal of approximately 90 degrees forming a shape which resembles the letter T.

Filter Lens: Optical plate used to keep harmful rays from reaching the eyes.

Flame-Cutting: A cutting process whereby a high temperature created by an oxygen-fuel torch is used to oxidize the metal being cut. A separate oxygen orifice in the torch tip carries in the oxygen under high pressure which facilitiates the oxidation and helps remove the oxidized material.

Flowmeter: A gas regulator used for inert gases which is measured in cubic feet per minute.

Flux: A cleansing agent used to chemically clean the base metal before soldering or brazing.

Forehand Welding: Welding in the direction in which the torch is pointed.

Forge Welding: An outdated method of welding by heating and hammering two pieces of metal together until they become one.

Fusible Plugs: A device used in conjunction with acetylene cylinders and designed to melt at 220 degrees Fahrenheit and relieve undue pressure.

Fusion: Complete union of two pieces of metal through melting.

Gas Cup: See NOZZLE

Gage: Device used to measure pressure in a regulator.

Ground Clamp: A spring clamp used to attach ground cable to a workpiece.

Hardfacing: The process of hardening a surface by applying a layer of wear-resistant alloy to the base metal.

Hardness: A material's ability to resist penetration.

Helium: An inert gas sometimes used as a substitute for Argon.

Helmet: A protective device used to cover the face, eyes, and neck from heat, spatter, and harmful rays during arc welding.

Hose: Flexible material used to transport gases from the regulators to the torch.

Igniter Current: The high frequency current used to help jump the arc and sustain the arc in AC TIG welding.

Impact Value: The amount of energy required to break a specimen with a fracturing blow.

Incandescent: The point at which a material becomes hot enough to give off light.

Inert Gas: A gas which will not support a chemical reaction when exposed to intense heat or other materials.

Infrared Rays: Heat rays produced by welding which require proper shielding to prevent eye injury.

Ingot: Large rectangular forms into which refined iron is poured after leaving the furnace.

Injector-Type Torch: High pressure oxygen is injected into the low pressure acetylene to draw it out of the torch.

Inner Cone: The blue conical shaped flame approximately 1/4" to 3/8" long which starts at the tip of the torch.

Izod Test: An impact test in which only one end of the specimen is held.

Lamellar: A platelike appearance.

Land: The small flat edge which results from the grinding of a chamfer on the parent metal. Also known as the root face.

Lap Weld: A triangular bead used to join two overlapping pieces of metal.

Limestone: A stone material used as a flux in the refinement of iron.

Linear Expansion: The amount in millionths of an inch that a specimen will enlarge per inch of length per degree Celsius rise in temperature.

Longitudinal Angle: The angle of the torch or electrode along the line of the weld.

Low-Alloy Steel: A form of steel containing a small percentage of manganese, nickel, molybdenum, or chrome.

MIG Carbon Dioxide: The MIG welding process using CO_2 when welding carbon and low-alloy steel to produce deeper penetration.

Malleableizing: A heat treating process whereby white cast iron is annealed to increase its ductility.

Martensite: The result of the quenching of austenite. It is an extremely hard and brittle form of iron.

Medium-Pressure Torch: Both gases are delivered at the same pressure and mixed equally.

Metallurgy: The study of metals and their properties.

Mill Scale: A black scale of magnetic oxide of iron produced by the heating of iron.

Mixing Head: Combines the two gases into a usable fuel.

Neutral Flame: The resulting flame from the combustion of equal amounts of oxygen and acetylene. It is attained by making the acetylene feather the same length as the inner cone.

Nitriding: A form of case-hardening by absorbing nitrogen into the outer layers.

Normalizing: A method of slightly hardening steel without making it overly brittle.

Nozzle: Ceramic or Pyrex® device used to direct the gas to properly shield the arc in TIG and MIG welding.

Open-Hearth Furnace: A steel refining furnace which exposes the hearth of the furnace to the flames used to melt the iron.

Orifice: A small opening through which gases flow.

Oscillate: To move back and forth or side-to-side.

Oxidizing Flame: The resulting flame which contains an excess amount of oxygen. This flame can be identified by its short, pointed cone.

Oxyacetylene Welding: A welding process using the intensely hot flame produced by the combustion of the fuel derived from the mixture of oxygen and acetylene.

Oxygen: Colorless, odorless, tasteless gas slightly lighter than air.

Pack-Hardening: The hardening of steel by submerging it in carbon and heating it in a furnace.

Parent Metal: The metal to be welded, soldered, brazed, or cut.

Pearlite: A mechanical mixture of ferrite and cementite.

Penetration: The depth of fusion into the parent metal as measured from the surface of the parent metal.

Pig Iron: Crude iron which is the direct product of, and poured from, the blast furnace.

Plate: Any metal over 3/16" thick and 6" wide.

Polarity: The positive or negative characteristics of an electrical circuit.

Porosity: The gas pockets contained within a weld.

Preheat: The application of heat prior to welding to minimize the loss of heat to cold areas due to heat conduction.

Puddle: A concave pool of molten metal near the heat source.

Rectifier: An electrical device used to convert AC current into a pulsating DC current.

Regulator: Reduces high cylinder pressures to usable working pressures and keep the working pressure constant.

Residual Stress: The internal pressures which remain after heating and cooling metals.

Rockwell B-Scale: Scale for softer materials tested with a steel ball.

Rockwell C-Scale: Scale for harder materials tested with a diamond brale.

Rockwell Hardness Test: Tests materials for hardness by forcing either a steel ball or a diamond brale into the surface at 100 kilograms.

Root Opening: The distance between two pieces to be welded at the root of the joint.

Safety Cap: A removable metal cap used to protect the cylinder valve of an oxygen cylinder.

Scleroscope: A device used to measure the hardness of a material.

Shielding Gas: The gas used in TIG and MIG welding to shield the arc and the metal from the atmosphere and, therefore, from oxidation.

Short-Arc Welding: A MIG welding process using reduced heat and a pin arc for welding lightweight material.

Slag: Oxide impurities left on the surface of the metal after welding.

Slag Inclusions: Small nonmetallic oxides included in the weldment.

Soapstone: A soft stone used to mark guidelines on metal.

Soldering: Joining method whereby a nonferrous alloy flows between the two parts to be joined by capillary action. This process is done at less than 800 degrees Fahrenheit.

Sparklighter: A mechanical device used to create a spark and light an oxyacetylene torch.

Spark Testing: A method of identifying steel by examining the spark given off when the material is in contact with a grinding wheel.

Spatter: Small droplets of metal which spray in the area of an arc weld.

Specific Heat: The amount of heat needed to raise the temperature of an object one degree compared to the amount of heat required to raise the temperature of an equal mass of water one degree.

Specimen: A shape which meets exacting standards to be used for controlled testing.

Spectroscope: An optical device used to create and record the spectra of light given off by various metals.

Spheroidizing: Heating steel until the cementite assumes a globular form, this produces softness and good machinability.

Spray-Arc Welding: A MIG welding process where the electrode is melted in the gun and transferred to the parent metal in small globules.

Stickout: The distance the electrode extends beyond the gas nozzle in MIG and TIG welding.

Stress Relieving: The removal of the internal pressures created by heat treating processes.

Stringer Bead: An arc welding bead which is formed in a straight line with no side-to-side movement.

Tack Weld: A small weld used to hold the work in position and control distortion in the welding process.

Tee Weld: See FILLET WELD

Tempering: Heat treating steel to make it strong, tough, and usable without being excessively brittle.

Tensile Strength: A material's resistance to being pulled apart.

Thermal Conductivity: The ability to transfer heat from one area to another.

Thermal Expansion: The ability of a material to expand when subjected to varying amounts of heat.

Thorium: Metal coating added to tungsten electrodes to improve penetration.

Tip: The end of the oxyacetylene torch where the fuel burns.

Toe of Weld: The point where the face of a weld contacts the surface of the parent metal.

Torch: The device used to produce the heat necessary to fuse metals. May be of the gas type as in oxyacetylene welding or the arc-type as in TIG welding.

Transformer: An electrical device used to vary the voltage and amperage through the use of the electromagnetic induction.

Transverse Angle: The angle of the torch or electrode across the line of the weld.

Tungsten: A gray-white, heavy, hard metal with a high melting point used for electrodes in tungsten inert gas welding.

Undercutting: The thinning of the parent metal due to the melting and lack of filler metal in the vertical portion of the weld.

Ultraviolet Rays: High energy waves produced by arc welding which require shielding due to their intensity and ability to cause blindness.

Voltage: The amount of electromotive force applied to the arc welding circuit.

Weave Bead: An arc welding bead which is formed by using side-to-side movements while making the pass which results in a bead approximately three times the width of the electrode.

Welding Rod: Filler wire which is melted and fused into the parent metal.

Whiskers: Short lengths of electrode which have not been consumed into the MIG welding bead.

Wire Electrode: The consumable filler metal used in MIG welding.

Yield Point: The point at which a material will continue to stretch although no additional pulling force is applied.

OXYACETYLENE FLAME TRAITS

FOR WELDING

Acetylene Burning in Atmosphere
Open fuel gas valve until smoke clears from flame.

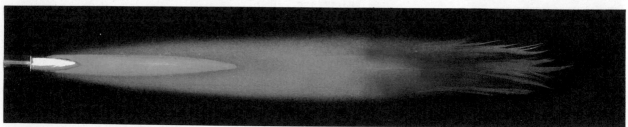

Carburizing Flame
(Excess acetylene with oxygen.) Used for hard-facing and welding white metal.

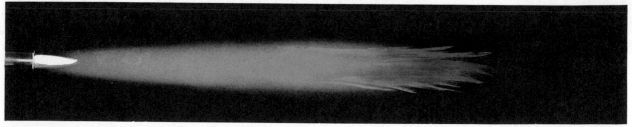

Neutral Flame
(Acetylene and Oxygen.) Temperature 6300 °F. For fusion welding of steel and cast iron.

Oxidizing Flame
(Acetylene and Excess Oxygen.) For braze welding with bronze rod.

These natural, unretouched color photographs give the operator an accurate guide for adjusting the oxyacetylene flame. Correct flame adjustment is so vital to successful welding and cutting that the best manipulative skill is lost unless the flame adjustment is correct.

FOR CUTTING

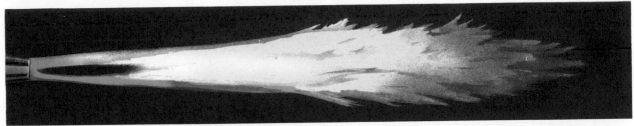

Acetylene Burning in Atmosphere
Open fuel gas valve until smoke clears from flame.

Carburizing Flame
(Excess acetylene with oxygen.) Preheat flames require more oxygen.

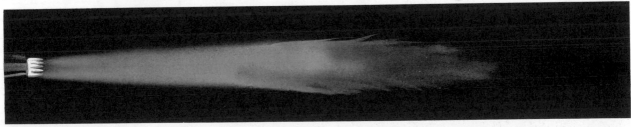

Neutral Flame
(Acetylene with Oxygen.) Temperature 6300 °F. Proper preheat adjustment for all cutting.

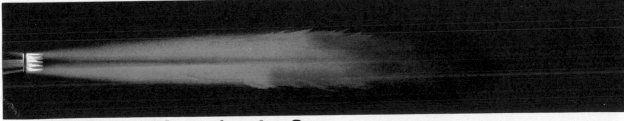

Neutral Flame with Cutting Jet Open
Cutting jet must be straight and clear.

Oxidizing Flame
(Acetylene with Excess Oxygen.) Not recommended for average cutting.